W9-BZB-122

2ND EDITION

ENERGY

IN THE 21ST CENTURY

2ND EDITION

ENERGY

IN THE 21ST CENTURY

JOHN R. FANCHI

Texas Christian University, USA

WITH CHRISTOPER J. FANCHI

World Scientific

NEW JERSEY • LONDON • SINGAPORE • BEIJING • SHANGHAI • HONG KONG • TAIPEI • CHENNAI

Published by

World Scientific Publishing Co. Pte. Ltd.
5 Toh Tuck Link, Singapore 596224
USA office: 27 Warren Street, Suite 401-402, Hackensack, NJ 07601
UK office: 57 Shelton Street, Covent Garden, London WC2H 9HE

British Library Cataloguing-in-Publication Data
A catalogue record for this book is available from the British Library.

ENERGY IN THE 21ST CENTURY (2nd Edition)

Desk Editor: Tjan Kwang Wei

ISBN-13 978-981-4322-04-1
ISBN-10 981-4322-04-0
ISBN-13 978-981-4324-54-0 (pbk)
ISBN-10 981-4324-54-X (pbk)

Printed in Singapore by Mainland Press Pte Ltd.

To the pioneers in the emerging energy industry –
for the benefit of future generations.

PREFACE TO THE SECOND EDITION

Many events that affect global energy production and consumption have occurred since the first edition of this book appeared in 2005. For example, the price of oil has ranged between US$30 per barrel and US$150 per barrel, and the demand for energy has been affected by a worldwide economic recession. The second edition updates data and expands material based on recent events.

The first edition of this book has been used at the text in an introductory energy course for a general college student population. This experience motivated the reorganization and addition of topical material, and inclusion of activities to enhance the value of the second edition as a textbook.

Most of the statistics presented in the book are from either the United States Energy Information Administration (www.eia.doe.gov) or the International Energy Agency (www.iea.org). The IEA is supported by industrialized nations, including the United States. The U.S. EIA is based in Washington, D.C., and the IEA is based in Paris, France. These data sources have been selected because they are widely used by policy makers.

We want to thank students, guest speakers, and colleagues in academia and industry for their comments and suggestions. Kathy Fanchi was again instrumental in preparing this book for publication.

John R. Fanchi
August 2010

PREFACE TO THE FIRST EDITION

My interest in energy began in the 1970's when I obtained degrees in physics from the Universities of Denver (B.S.), Mississippi (M.S.), and Houston (Ph.D.). I did some work in geothermal storage of solar energy as a post-doc in 1978, and then spent many years in the energy industry helping develop oil and gas reservoirs. I became a full time academic in 1998 when I joined the faculty of the Colorado School of Mines as a professor of petroleum engineering.

In the transition from industry to academia, I wanted to find out how long a college graduate today could expect to continue a career in the extraction of fossil fuels. After studying several forecasts of energy production, I was convinced that fossil fuels would continue to be an important part of the energy mix while other energy sources would increase in importance. To help prepare students to function as energy professionals, I developed an energy course at the Colorado School of Mines and published the textbook **Energy: Technology and Directions for the Future** (Elsevier – Academic Press, Boston, 2004).

I realized as I was developing the energy course that much of the material in the textbook is suitable for a general audience. This book, **Energy in the 21st Century**, is a non-technical version of **Energy: Technology and Directions for the Future**. **Energy in the 21st Century** was written to give the concerned citizen enough information about energy to make informed decisions. Readers who would like more detailed information or a more complete list of references should consult the textbook **Energy: Technology and Directions for the Future**.

I want to thank my students and guest speakers for their comments during the preparation of my energy course. Tony Fanchi helped prepare many of the figures in the book, and Kathy Fanchi was instrumental in the preparation and production of the book. Even though there are many more topics that could be discussed, the material in **Energy in the 21st Century** should expose you to a broad range of energy types and help you develop an appreciation of the role that each energy type may play in the future.

John R. Fanchi

CONTENTS

ABOUT THE AUTHORS

John R. Fanchi is a Professor in the Department of Engineering and Energy Institute at Texas Christian University, Fort Worth, Texas where he teaches courses in energy and engineering. Previously, Fanchi taught petroleum and energy engineering courses at the Colorado School of Mines, worked in the technology centers of four energy companies, and has been President of the International Association for Relativistic Dynamics. He has a Ph.D. in physics from the University of Houston and is the author of several books in the areas of energy, physics, engineering, earth science, and mathematics. His books include **Integrated Reservoir Asset Management** (Elsevier, 2010), **Energy in the 21st Century** (World Scientific, 2005), **Energy: Technology and Directions for the Future** (Elsevier-Academic Press, 2004), **Principles of Applied Reservoir Simulation, 3rd Edition** (Elsevier, 2006), **Math Refresher for Scientists and Engineers, 3rd Edition** (Wiley, 2006), **Shared Earth Modeling** (Elsevier, 2002), **Integrated Flow Modeling** (Elsevier, 2000), and **Parametrized Relativistic Quantum Theory** (Kluwer, 1993). He co-edited the General Engineering volume of the **Petroleum Engineering Handbook** (SPE, 2006, Volume 1).

Christopher J. Fanchi has a B.A. in Business Administration and Economics from Colorado State University. He has worked for Fanchi Enterprises since 2000, including contributions to **Integrated Reservoir Asset Management** (Elsevier, 2010), **Math Refresher for Scientists and Engineers, 3rd Edition** (Wiley, 2006) and **Integrated Flow Modeling** (Elsevier, 2000).

CHAPTER 1

A BRIEF HISTORY OF ENERGY CONSUMPTION

We all make decisions about energy. We decide how much electricity we will use to heat or cool our homes. We decide how far we will go every day and the mode of transportation we will use. Those of us in democracies choose leaders who create budgets that can support new energy initiatives or maintain a military capable of defending energy supply lines. Each of these decisions and many others impacts the global consumption of energy and the demand for available natural resources. The purpose of this book is to give you the information you need to help you make informed decisions.

The choices we make today will affect generations to come. What kind of future do we want to prepare for them? What kind of future is possible? We can make the best decisions by being aware of our options and the consequences of our choices. In this book, we consider the location, quantity and accessibility of energy sources. We discuss ways to distribute available energy, and examine how our choices will affect the economy, society, and the environment. Our understanding of each of these issues will help us on our journey to energy independence. We begin by defining energy and reviewing our history of energy consumption.

1.1 WHAT IS ENERGY?

Energy is the ability to do work. It can be classified as stored (potential) energy, and working (kinetic) energy. Potential energy is the ability to produce motion, and kinetic energy is the energy of motion. Forms of energy include energy of motion (kinetic energy), heat (thermal energy), light (radiant energy), photosynthesis (biological energy), stored energy in a battery (chemical energy), stored energy in a capacitor (electrical energy), stored energy in a nucleus (nuclear energy), and stored energy in a gravitational field (gravitational energy).

Sources of energy with some common examples include biomass (firewood), fossil fuels (coal, oil, natural gas), flowing water (hydroelectric dams), nuclear materials (uranium), sunlight, and geothermal heat (geysers). Energy sources may be classified as renewable or non-renewable. Non-renewable energy is energy that is obtained from sources at a rate that exceeds the rate at which the sources are replenished. Examples of non-renewable energy sources include fossil fuels and nuclear fission material such as uranium. Renewable energy is energy that is obtained from sources at a rate that is less than or equal to the rate at which the sources are replenished. Examples of renewable energy include solar energy and wind energy.

Renewable and non-renewable energy sources are considered primary energy sources because they provide energy directly from raw fuels. A fuel is a material which contains one form of energy that can be transformed into another form of energy. Primary energy is energy that has not been obtained by anthropogenic conversion or transformation. The term "anthropogenic" refers to human activity or human influence. Primary energy is often converted to secondary energy for more convenient use in human systems. Hydrogen and electricity are considered secondary sources of energy, or "carriers" of energy. Secondary energy sources are produced from primary sources of energy. Secondary sources of energy can store and deliver energy in a useful form.

Modern civilization depends on the observation that energy can change from one form to another. If you hold this book motionless above a table and then release it, the book will fall onto the table. The book has potential energy when it is being held above the table. The potential energy is energy associated with the position of the book in a gravitational field. When you drop the book, the energy of position is transformed into energy of motion, or kinetic energy. When the book hits the table, some of the kinetic energy is transformed into sound (sonic energy), and the rest of the kinetic energy is transformed into energy of position (potential energy) when the book rests on the table top.

Energy transformation is needed to produce commercial energy. As an illustration, suppose we consider a coal-fired power plant. Coal stores energy as chemical energy. Combustion, or burning the coal, transforms chemical energy into heat energy. In steam power plants, the heat energy changes water into steam and increases the energy of motion, or kinetic energy, of the steam. Flowing steam spins a turbine in a generator. The mechanical energy of the spinning turbine is converted to electrical energy in the generator. In a real system, energy is lost so that the efficiency of electrical energy generation from the combustion of coal is less than 100%. A measure of the energy that is available for doing useful work is called exergy.

Real power systems transform energy into useful work, but some of the energy is wasted. The energy efficiency of a system is the amount of energy needed by the system to perform a specific function divided by the amount of energy that is supplied to the system. Energy efficiency has a value between 0% and 100%. Some of the energy supplied to a real system is lost as non-useful energy so that the energy efficiency is less than 100%. For example, suppose we have two light bulbs A and B. Both light bulbs provide the same amount of light, but light bulb B uses less energy than light bulb A because light bulb B produces less heat than light bulb A. Light bulb B has a higher energy efficiency than light bulb A because light bulb B uses less energy to achieve its intended purpose, to provide light.

In the light bulb example, we can reduce energy use by adopting a more energy efficient technology. Another way to reduce energy use is to turn off the light when it is not needed. In this case, we are conserving energy by changing our behavior. Energy conservation is achieved by adopting a behavior that results in the use of less energy. An improvement in energy efficiency or conservation can be viewed as increasing energy supply because improving energy efficiency or conservation lets us get more value from existing energy sources.

Point to Ponder: What does an energy unit mean to me?
To get an idea of the meaning of an energy unit such as kilocalorie or megajoule, it is helpful to compare the energy consumed by the operation of modern devices. For example, a 1200 Watt hair dryer uses approximately one megajoule of energy in 15 minutes. A megajoule is 1 million Joules, which can be written as 10^6 Joules or 10^6 J. A 100 Watt light bulb uses approximately one megajoule of energy in about three hours. [Fanchi, 2004, Exercise 1-3]

If we run the 1200 Watt hair dryer for one hour, we will use 1.2 kilowatt-hours of energy. We abbreviate 1.2 kilowatt-hours as 1.2 kWh. One kWh equals 1 kW times 1 hr, or about 3.6×10^6 J of energy. A typical American household will use between 20 and 50 kWh per day. Energy usage depends on many factors, such as use of appliances, heating or cooling, etc.

A typical power plant provides approximately 1000 megawatts of power, which is abbreviated as 1000 MW of power. The power plant can provide power to approximately 900,000 households that use 10,000 kWh per year for each household.

A unit of energy that is commonly used for discussing energy on a national scale is the quad. One quad equals one quadrillion British Thermal Units (BTU) or 10^{15} BTU. A BTU is approximately 1000 Joules, so one quad is approximately 10^{18} Joules. A quad is comparable in magnitude to global energy values. For example, in 2006 the United States consumed about 100 quads of energy and the world consumed about 472 quads of energy.

Units and scientific notation are reviewed in Appendix A.

1.2 Historical Energy Consumption

The history of energy consumption shows how important energy is to the quality of life for each of us. Societies have depended on different types of energy in the past, and societies have been forced to change from one energy type to another. Global energy consumption can be put in perspective by considering the amount of energy consumed by individuals.

E. Cook [1971] provided estimates of daily human energy consumption at six different periods of societal development. The six periods from oldest to most recent are the Primitive Period, the Hunting Period, the Primitive Agricultural Period, the Advanced Agricultural Period, the Industrial Period, and the Technological Period. Cook's estimates are given in Table 1-1 for each period. The table shows that personal energy consumption was relatively constant until the Advanced Agricultural period when it increased substantially.

Table 1-1
Historical Energy Consumption [Cook, 1971]

Period	Era	Daily per capita Consumption (1000 kcal)				
		Food	H & C*	I & A**	Trans.***	Total
Primitive	1 million B.C.	2				2
Hunting	100,000 B.C	3	2			5
Primitive Agricultural	5000 B.C.	4	4	4		12
Advanced Agricultural	1400	6	12	7	1	26
Industrial	1875	7	32	24	14	77
Technological	1970	10	66	91	63	230

* H & C = Home and Commerce
** I & A = Industry and Agriculture
*** Trans. = Transportation

Energy is essential for life, and food was the first source of energy. Cook assumed the only source of energy consumed by a person living during the period labeled "Primitive" was food. Cook's energy estimate was for an East African about one million years ago. Humans require approximately 2000 kilocalories (about eight megajoules) of food per day. One food Calorie is equal to one kilocalorie, or 1000 calories. One calorie is the amount of energy required to raise the temperature of one gram of water one degree Centigrade. A change in temperature of one degree Centigrade is equal to a change in temperature of 1.8 degrees Fahrenheit.

The ability to control fire during the Hunting period let people use wood to heat and cook. Fire provided light at night and could illuminate caves. Firewood was the first source of energy for consumption in a residential setting. Cook's estimate of the daily per capita energy consumption for Europeans about 100,000 years ago was 5,000 kilocalories (about 21 megajoules).

Figure 1-1. Animal Labor in Ahmedabad, India (Fanchi, 2000)

The Primitive Agricultural period was characterized by the domestication of animals. Humans were able to use animals to help them grow crops and cultivate their fields. The ability to grow more food than you needed became the impetus for creating an agricultural industry. Cook's estimate of the daily per capita energy consumption for people in the

Fertile Crescent circa 5000 B.C. was 12,000 kilocalories (about 50 megajoules). Humans continue to use animals to perform work (Figure 1-1).

More energy was consumed during the Advanced Agricultural period when people learned to use coal, and built machines to harvest the wind and water. By the early Renaissance, people were using wind to push sailing ships, water to drive mills, and wood and coal for generating heat. Transportation became a significant component of energy consumption by humans. Cook's estimate of the daily per capita energy consumption for people in northwestern Europe circa 1400 was 26,000 kilocalories (about 109 megajoules).

The steam engine ushered in the Industrial period. It provided a means of transforming heat energy to mechanical energy. Wood was the first source of energy for generating steam in steam engines. Coal, a fossil fuel, eventually replaced wood and hay as the primary energy source in industrialized nations. Coal was easier to store and transport than wood and hay, which are bulky and awkward. Coal was useful as a fuel source for large vehicles, such as trains and ships, but of limited use for personal transportation. Oil, another fossil fuel, was a liquid and contained about the same amount of energy per unit mass as coal. Oil could flow through pipelines and tanks. People just needed a machine to convert the energy in oil to a more useful form. Cook's estimate of the daily per capita energy consumption for people in England circa 1875 was 77,000 kilocalories (about 322 megajoules).

The modern Technological period is associated with the development of internal combustion engines, and applications of electricity. Internal combustion engines can vary widely in size and use oil. The internal combustion engine could be scaled to fit on a wagon and create "horseless carriages." The transportation system in use today evolved as a result of the development of internal combustion engines. Electricity, by contrast, is generated from primary energy sources such as fossil fuels. Electricity generation and distribution systems made the widespread use of electric motors and electric lights possible. One advantage of electricity as an energy source is that it can be transported easily, but electricity is difficult to store. Cook's estimate of the daily per capita energy consumption for people in the United States circa 1970 was 230,000 kilocalories (about 962 megajoules).

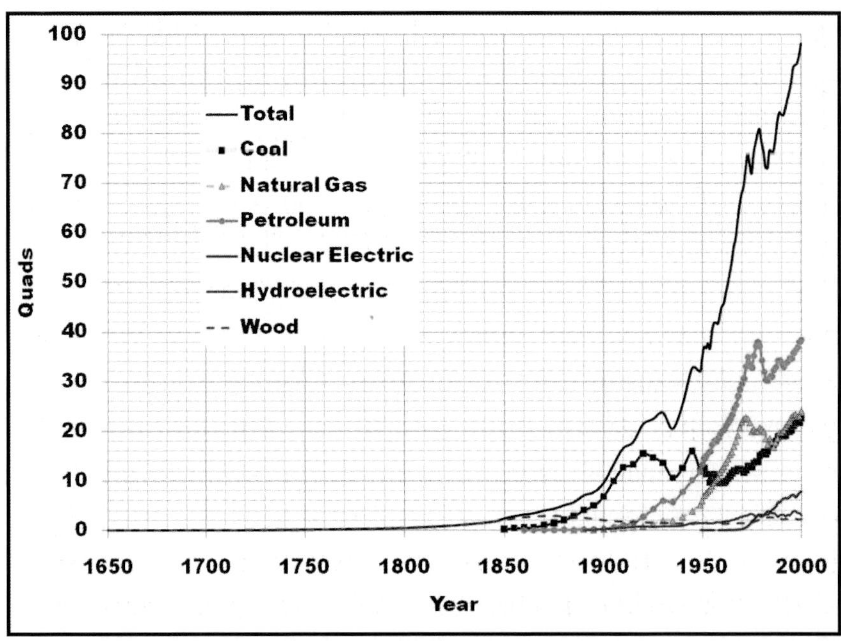

Figure 1-2. Historical Energy Consumption in the United States [US EIA website, 2001]

Figure 1-2 shows the consumption of energy in the United States from 1650 to 2000. The figure clearly shows that wood was the primary energy source for most of history. The transition from wood (a renewable energy source) to fossil fuels (a non-renewable energy source) began in the middle of the 19th century. Fossil fuels became the dominant energy source from the mid-19th century through the end of the 20th century. A similar scenario applies to other developed nations.

Table 1-2 shows that approximately 74 quads of energy were produced and 99 quads of energy were consumed by the United States in 2008. The energy that was not produced in the United States was imported.

Table 1-2
2008 United States Energy Production and Consumption
[US EIA website, 2008]

Energy Source	Production (quads)	Consumption (quads)
Total	73.71	99.30
Fossil Fuels	57.94	83.44
Electricity Net Imports		0.11
Nuclear Electric Power	8.46	8.46
Renewable Energy	7.32	7.30

The data in Table 1-2 is taken from data published by the Energy Information Administration (EIA) of the United States Department of Energy. Additional historical data for the United States from the EIA is presented in Appendix B, and analogous data for the world is presented in Appendix C. Energy data for the United States is used to illustrate energy production and consumption for an individual nation. All of the data reported in this book is subject to revision as data collection agencies acquire revisions and corrections of data from institutions to monitor and report energy consumption and production. Although the data is subject to revision, it is useful to view actual reported data so that we can get an idea of the magnitude of energy needed and the relative contributions of different energy sources to the energy mix.

Table 1-2 shows the production and consumption of energy in the United States. The unit of energy is the quad, or quadrillion BTU. Although four energy sources are identified, the three dominant energy sources are fossil fuels, nuclear electric power, and renewable energy.

Table 1-3 presents the relative contributions of different energy sources to 2008 United States energy production and consumption as percent of total values shown in Table 1-2. Fossil fuels were the dominant contributor to the United States energy mix in 2008. This provides a snapshot of the United States energy mix. The data in Appendix B shows that the relative contribution of energy sources to the United States energy mix is changing with the contribution of fossil fuels declining while the contribution of renewable energy sources is increasing. The increasing role of renewable energy sources is also occurring on a global scale, as discussed in Section 1-5.

Table 1-3
2008 United States Energy Production and Consumption as % of Annual Amount
[US EIA website, 2008]

Energy Source	Production (%)	Consumption (%)
Total	100.0	100.0
Fossil Fuels	78.6	84.0
Electricity Net Imports		0.1
Nuclear Electric Power	11.5	8.5
Renewable Energy	9.9	7.4

Table 1-4 shows how the energy consumed in the United States in 2008 was distributed between energy sectors. Definitions of the composition of each energy sector are provided by the United States EIA. The transportation sector includes vehicles that transport people or goods.

Table 1-4
2008 United States Energy Consumption by Sector
[US EIA website, 2008]

Sector	Consumption (%)
Transportation	27.8
Industrial	20.6
Residential and Commercial	10.8
Electric power	40.1

1.3 Energy Consumption and the Quality of Life

The existing per capita energy consumption discussed above gives us an idea of how much energy each of us uses today, but it does not tell us how much energy each of us should use. One way to estimate the amount of energy each person should use is to examine the relationship between energy consumption and quality of life.

Quality of life is a subjective concept that can be quantified in several ways. The United Nations calculates a quantity called the Human Development Index (HDI) to provide a quantitative measure of the quality of life. The HDI measures human development in a country using three basic factors: health, knowledge, and standard of living. Table 1-5 presents measures for each factor.

Gross Domestic Product (GDP) accounts for the total output of goods and services from a nation and is a measure of the economic growth of the nation. The HDI is a fraction that varies from zero to one. A value of HDI that approaches zero is considered a relatively low quality of life, while a value of HDI that approaches one is considered a high quality of life.

Table 1-5
Basic Factors of the United Nations Human Development Index

Factor	Measure
Health	Life expectancy at birth
Knowledge	Combination of adult literacy and school enrollment
Standard of Living	GDP per capita

A plot of HDI versus per capita electricity consumption for all nations with a population of at least one million people is shown in Figure 1-3. Per capita electricity consumption is the total amount of electricity consumed by the nation divided by the population of the nation. The amount of electricity is expressed in kilowatt-hours (kWh) in the figure. Per capita electricity generation is an estimate of the average amount of electricity consumed by each individual in the nation. The calculation of per capita electricity consumption establishes a common basis for comparing the consumption of electricity between nations with large populations and nations with small populations. The HDI data are 2006 data from the United Nations Human Development Report [UN HDI, 2009], and annual per capita electricity consumption data are 2006 data reported by the Energy Information Administration of the United States Department of Energy [US EIA website, 2009].

Figure 1-3 shows that quality of life, as measured by HDI, increases as per capita electricity consumption increases. It also shows that the increase is not linear; the improvement in quality of life begins to level off when per capita electricity consumption rises to about 4000 kilowatt-hours. A similar plot can be prepared for per capita energy consumption and is shown in Figure 1-4.

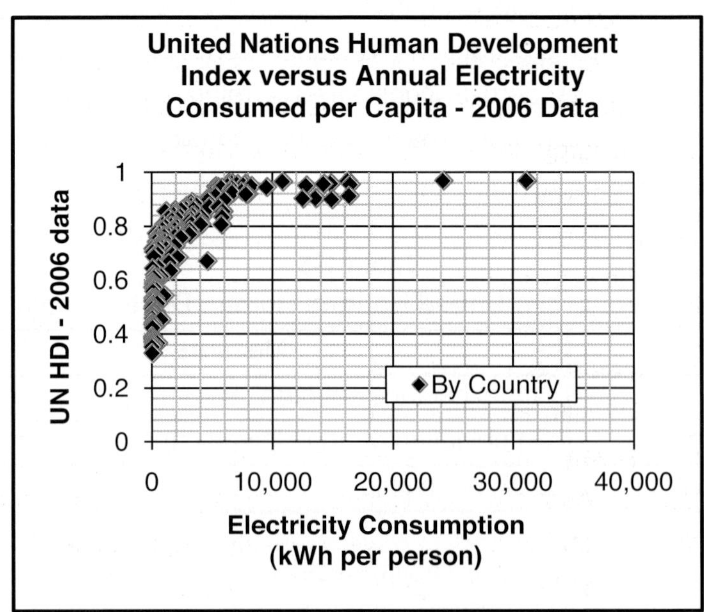

Figure 1-3. Human Development and Annual Electricity Consumption

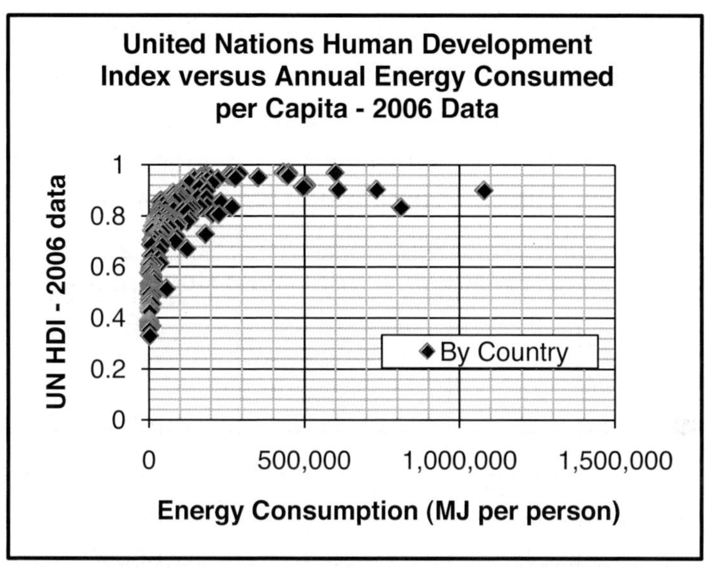

Figure 1-4. Human Development and Annual Energy Consumption

Figure 1-4 is a plot of HDI versus per capita energy consumption for nations with a population of at least one million people. The HDI data are 2006 data from the United Nations Human Development Report [UN HDI, 2009], and annual per capita energy consumption data are 2006 data reported by the Energy Information Administration of the United States Department of Energy [US EIA website, 2009]. The figure shows that quality of life increases as per capita energy consumption increases. As in Figure 1-3, the increase is not linear; the improvement in quality of life begins to level off when per capita energy consumption rises to about 200,000 megajoules per person.

The countries with the largest HDI values, in excess of 90%, are nations with relatively mature economies such as western European nations, Canada, Australia, the United Kingdom, Japan, and the United States. These countries tend to have relatively large middle classes. Table 1-6 lists per capita consumption for the 15 countries with the largest UN HDI. The numbers are subject to change when databases are updated or corrected, but the data in the table does illustrate the relative magnitude of per capita consumption for each country. Some people may be surprised to see that the United States is not the largest consumer of energy or electricity on a per capita basis.

Table 1-6
Per Capita Consumption for Countries with the Largest UN HDI (2006 Data)

Country	UN HDI	Per Capita Consumption	
		Electricity (kWh/person)	Energy (MJ/person)
Iceland	0.968	31103	599,905
Norway	0.968	24175	433,352
Canada	0.967	16228	450,670
Australia	0.965	10855	292,118
Ireland	0.960	6320	182,929
Netherlands	0.958	6643	264,681
Sweden	0.958	14812	259,274
Japan	0.956	7705	188,519
Luxembourg	0.956	14224	447,457
France	0.955	7063	190,669
Switzerland	0.955	7811	180,055
Finland	0.954	16446	266,622
Denmark	0.952	6362	170,199
Austria	0.951	7610	197,501
United States	0.950	12789	352,990

Point to Ponder: How can we use quality of life to forecast energy consumption?

The data used to prepare Figures 1-2 and 1-3 can also be used to make a quick forecast of energy demand. Suppose we assume that the world population will stabilize at approximately 8 billion people in the 21st century and that all people will want the quality of life represented by an HDI value of 0.9 (which is approximately the HDI value achieved by Portugal, Qatar, and the Czech Republic). In this scenario, the annual per capita energy demand from Figure 1-4 is approximately 200,000 megajoules per person, or 1.6×10^{15} MJ $\approx$ 1500 quads. The world population of approximately 6.4 billion people consumed approximately 472 quads of energy in 2006. According to this scenario, worldwide energy demand will need to triple by the end of the 21st century when compared to worldwide energy consumption in 2006. Annual per capita energy consumption will have to increase from an average of 76,000 megajoules per person in 2006 to the desired value of 200,000 megajoules per person in 2100. This calculation illustrates the types of assumptions that must be made to prepare forecasts of energy demand. At the very least, a forecast of demand for energy at the end of the 21st century needs to provide an estimate of the size of the population and the per capita demand for energy at that time. [Fanchi, 2004, Chapter 1]

1.4 Energy in Transition

Coal was the first fossil fuel to be used on a large scale. J.U. Nef [1977] described 16th century Britain as the first major economy in the world that relied on coal. Britain was dependent on wood before it switched to coal. The transition from wood to coal during the period from about 1550 A.D. to 1700 A.D. was made necessary by the excessive consumption of wood that was leading to the deforestation of Britain. Coal was a combustible fuel that could be used as an alternative to wood.

Coal was the fuel of choice during the Industrial Revolution. It was used to boil steam for steam turbines and steam engines. Coal was used in transportation to provide a combustible fuel for steam engines on trains and ships. The introduction of the internal combustion engine made it possible for oil to replace coal as a fuel for transportation. Coal is used today to provide fuel for many coal-fired power plants.

People have used oil for thousands of years [Yergin, 1992, Chapter 1]. Civilizations in the Middle East, such as Egypt and Mesopotamia, collected oil in small amounts from surface seepages as early as 3000-2000 B.C. During that period, oil was used in building construction, waterproofing boats and other structures, setting jewels, and mummification. Arabs began using oil to create incendiary weapons as early as 600 A.D. By the 1700's, small volumes of oil from sources such as surface seepages and mine shafts were being used in Europe for medicinal purposes and in kerosene lamps. Larger volumes of oil could have been used, but Europe lacked adequate drilling technology.

Pulitzer Prize winner Daniel Yergin [1992, page 20] chose George Bissell of the United States as the person most responsible for creating the modern oil industry. Bissell realized in 1854 that rock oil – as oil was called in the 19th century to differentiate it from vegetable oil and animal fat – could be used as an illuminant in lamps. He gathered a group of investors together in the mid-1850's. The group formed the Pennsylvania Rock Oil Company of Connecticut and selected James M. Townsend to be its president.

Bissell and Townsend knew that oil was sometimes produced along with water from water wells. They believed that rock oil could be found below the surface of the earth by drilling for oil in the same way that water wells were drilled. Townsend commissioned Edwin L. Drake to drill a well in Oil Creek, near Titusville, PA. The location had many oil seepages. The project began in 1857 and encountered many problems. By the time Drake struck oil on Aug. 27, 1859, a letter from Townsend was en route to Drake to inform him that funds were to be cut off [van Dyke, 1997].

Drake's well caused the value of oil to increase dramatically. Oil could be refined for use in lighting and for cooking. The substitution of rock oil for whale oil, which was growing scarce and expensive, reduced the need to hunt whales for fuel to burn in lamps. Within fifteen months of Drake's strike, Pennsylvania was producing 450,000 barrels of oil a year from seventy-five wells. By 1862, three million barrels of oil were being produced and the price of oil dropped to ten cents a barrel [Kraushaar and Ristinen, 1993].

The Pennsylvania oil fields provided a relatively small amount of oil to meet demand. In 1882, the invention of the electric light bulb caused a drop in the demand for kerosene. The drop in demand for rock oil was short lived, however. The quickly expanding automobile industry needed oil for fuel and lubrication. New sources of oil were discovered in the early 20th century. Oil was found in Ohio and Indiana, and later in the San Fernando Valley in California and near Beaumont, Texas.

Figure 1-5. Spindletop, Gladys City Boomtown Museum, Beaumont, Texas (Fanchi, 2003)

The world's first gusher, a well that produced as much as 75,000 barrels of oil per day, was drilled at Spindletop Hill near Beaumont (Figure 1-5). The well was named Lucas-1 after Anthony F. Lucas, its driller and an immigrant from the Dalmatian coast (now Croatia). The well was notable because it was drilled using rotary drilling, a modern drilling technique discussed in more detail later, and it produced so much oil at such a high rate that it demonstrated oil could be used as an energy source on a global scale.

Industrialist John D. Rockefeller began Standard Oil in 1870 and by 1879 the company held a virtual monopoly over oil refining and transportation in the United States. Rockefeller's control of the oil business made him rich and famous. The Sherman Antitrust Act of 1890 was used by the United States government to break Rockefeller's grip on the oil industry. Standard Oil was found guilty of restraining trade and a Federal court ordered the dissolution of Standard Oil in 1909. The ruling was upheld by the United States Supreme Court in 1911 [Yergin, 1992, Ch. 5].

By 1909, the United States produced more oil than all other countries combined, producing half a million barrels per day. Up until 1950, the United States produced more than half of the world's oil supply. Discoveries of large oil deposits in Central and South America and the Middle East led to decreased United States production. Production in the United States peaked in 1970 and has since been declining. However, oil demand in the United States and elsewhere in the world has continued to grow. Since 1948, the United States has imported more oil than it exports. Today, the United States imports about half of its oil needs.

Until 1973, oil prices were influenced by market demand and the supply of oil that was provided in large part by a group of oil companies called the "Seven Sisters." This group included Exxon, Royal Dutch/Shell, British Petroleum (BP), Texaco, Mobil, Standard Oil of California (which became Chevron), and Gulf Oil. In 1960, Saudi Arabia led the formation of the Organization of Petroleum Exporting Countries, commonly known as OPEC. It was in 1973 that OPEC became a major player in the oil business by raising prices on oil exported by its members. This rise in price contributed to the "first oil crisis" as prices for consumers in many countries jumped. OPEC members in 2008 are Algeria, Angola, Ecuador, Iran, Iraq, Kuwait, Libya, Nigeria, Qatar, Saudi Arabia, United Arab Emirates, and Venezuela [US EIA website, 2009, OPEC Revenues Factsheet].

Today, fossil fuels are still the primary fuels for generating electrical power, but society is becoming increasingly concerned about the global dependence on finite resources and the environmental impact of fossil fuel combustion. As a result, society is in the process of changing the global energy mix from an energy portfolio that is heavily dependent on fossil fuels to an energy portfolio that depends on several energy sources. The transition process began in the latter half of the 20th century and is illustrated in Figure 1-6.

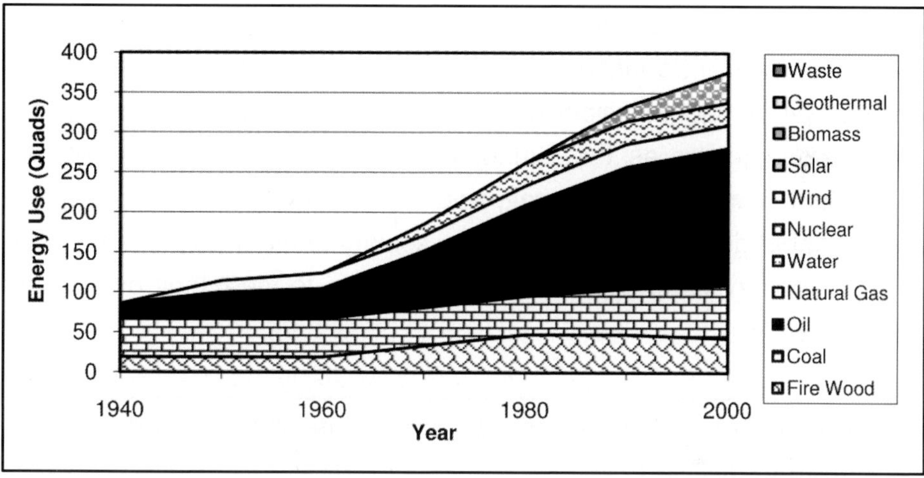

Figure 1-6. World Historical Energy Consumption

Figure 1-6 shows total energy consumption in quads from 1940 to 2000. In 1940, the world relied on firewood, coal and oil. Natural gas, energy from water, especially hydropower from dams, and nuclear energy joined firewood, coal and oil as important contributors to the energy mix by the end of the 20th century. Other energy sources – identified as wind, solar, biomass, geothermal and waste in Figure 1-6 – were beginning to make an appearance in the global energy mix at the beginning of the 21st century. They do not appear in the figure because their impact was negligible in the last half of the 20th century. One of the factors that supported the selection of fossil fuels and nuclear energy as fuels of choice is energy density.

Energy density is energy contained per unit volume of material. Fossil fuels have relatively large energy densities and have been preferentially chosen as the raw fuel for power plants. Raw fuels such as oil, coal, natural gas, and uranium are present in nature and can be used to provide primary energy.

The dominance of fossil fuels in the energy mix at the end of the 20th century is being replaced by a move toward sustainable energy. Sustainable energy is the mix of energy sources that will allow society to meet its present energy needs while preserving the ability of future generations to meet their needs. This definition is a variation of the concept of sustainable development introduced in 1987 in a report prepared by the United Nations' World Commission on Environment and Development. The Commission, known as the Brundtland Commission after chairwoman Gro Harlem Brundtland of Norway, said that society should adopt a policy of sustainable development that allows society to meet its present needs while preserving the ability of future generations to meet their own needs [WCED, 1987].

Point to Ponder: Why should I care about the global distribution of energy?

Suppose a country with a population of 20 million people wants to provide enough energy to sustain a quality of life corresponding to a United Nations HDI of 0.9. The country will require 200,000 megajoules per person of energy each year. This corresponds to approximately 127 power plants with 1000 megawatts capacity each. [Fanchi, 2004, Exercise 1-10] Where will this energy come from?

Today, energy on a national scale comes primarily from fossil fuels such as oil, gas and coal. In a few countries such as France, it is provided by nuclear fission. If the country does not have significant reserves of fossil fuels or uranium – a material needed for most nuclear fission reactors – it will have to import the materials it needs. In this case the country is a "have not" country that is dependent on countries that have the resources and technology it needs. This creates an opportunity for "have" countries to manipulate "have not" countries. On the other hand, it creates an incentive for "have not" countries to use its human resources to take what is needed. For example, the "have not" country could maintain a large standing army or sponsor acts of violence to influence "have" countries. The global distribution of energy influences relationships between nations and can affect geopolitical stability.

1.5 "DECARBONIZATION"

Energy forecasts rely on projections of historical trends. Table 1-7 presents historical data for global energy production reported by the United States Energy Information Administration for the last four decades of the 20th century.

Table 1-7
World Primary Energy Production by Source, 1970 – 2006
[US EIA website, 2009]

Primary Energy	Primary Energy Production (as % of total)				
	1970	1980	1990	2000	2006
Total	100.0	100.0	100.0	100.0	100.0
Fossil Fuels	93.2	90.1	86.7	85.4	86.2
Nuclear Electric Power	0.4	2.6	5.8	6.5	5.9
Renewable Energy	6.4	7.2	7.5	8.1	7.9
Total (quads)	215.4	287.6	349.9	395.7	469.4

The table shows historical energy production as a percent contribution to the total energy mix. The row of data labeled "Fossil Fuels" includes coal, petroleum and natural gas. Petroleum refers to hydrocarbon liquids such as crude oil and natural gas plant liquids. The row of data labeled "Renewable Energy" includes hydroelectric energy, geothermal energy, wind energy, solar energy, and bioenergy. The line of data labeled "Total (quads)" shows the total energy in quads produced for the specified year.

Published statistical data are subject to revision, even if the data are historical data that have been published by a credible source. Data revisions may change specific numbers as new information is received and used to update the database, but it is reasonable to expect the data presented in Table 1-7 to show qualitatively correct trends.

The data in Table 1-7 are graphically displayed in Figure 1-7. The data show the dominance of fossil fuels in the energy mix at the end of the 20th century. The current contribution from fossil fuels is approximately 86%.

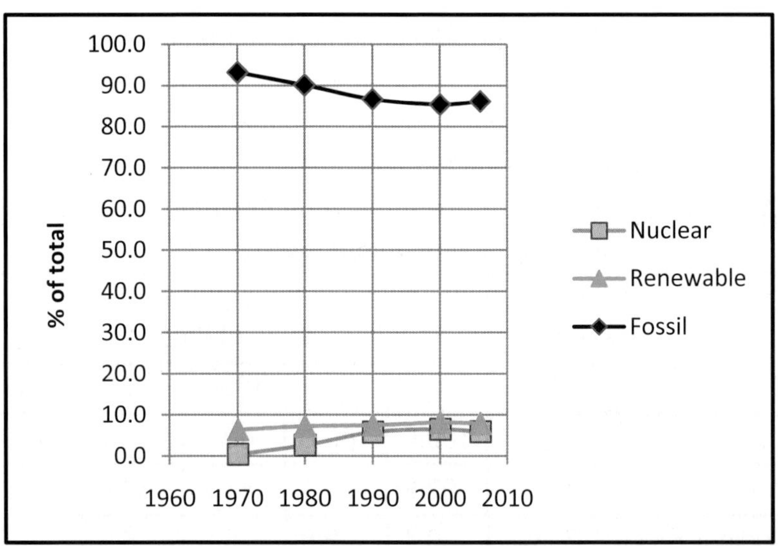

Figure 1-7. Percent Contribution of Different Energy Types to Historical Energy Production for the World from 1970 to 2006

The trend in the 20th century has been a "decarbonization" process, that is, a move away from fuels with many carbon atoms to fuels with few or no carbon atoms. H.J. Ausubel [2000, page 18] defined decarbonization as "the progressive reduction in the amount of carbon used to produce a given amount of energy." Figure 1-8 illustrates how the carbon to hydrogen ratio (C:H) declines as the fuel changes from carbon-rich coal to carbon-free hydrogen. The result of decarbonization will be a low carbon economy, or a low fossil fuel economy, but not necessarily a hydrogen economy. The use of hydrogen as a fuel in a hydrogen economy and other possible scenarios are discussed in more detail later.

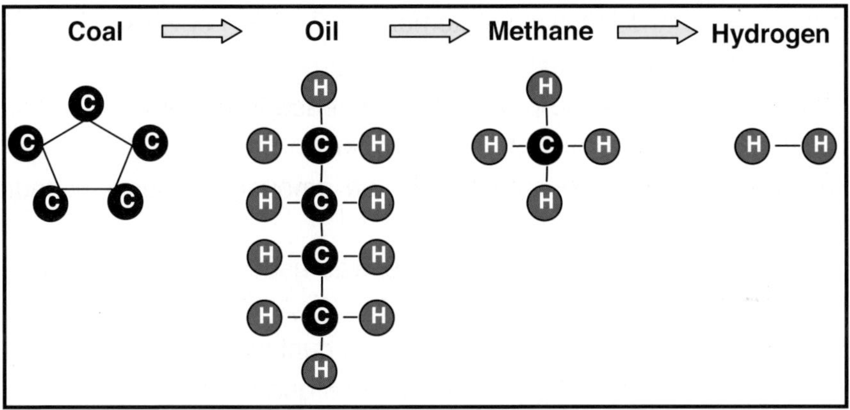

Figure 1-8. Decarbonization Among Fuels

Point to Ponder: Why would an oil rich country worry about alternative energy?

The trend toward decarbonization can help explain why a country, like Iran, that is rich in fossil fuels would seek to develop nuclear energy and alternative energy sources. Even if the country is able to export fossil fuels such as oil and gas for centuries to come at its current rate of export, the country can look at forecasts of energy consumption and see that the market is changing. They can use revenue from the sale of fossil fuels to help them transition to new energy sources. The remaining fossil fuels, especially oil, can be used for other applications besides fuel. For example, oil is used in the manufacture of plastics and other refined products such as lubricants.

1.6 Activities

True-False

Specify if each of the following statements is True (T) or False (F).

1. The steam engine transformed heat energy to mechanical energy.
2. Decarbonization is the progressive increase in the amount of carbon used to produce a given amount of energy.
3. Per capita energy consumption is the amount of energy consumed per person.
4. The Sherman Antitrust Act of 1890 was used to break up Standard Oil in 1909.
5. The use of rock oil reduced the need to hunt whales for fuel.
6. A quad is a unit of energy and refers to a quadrillion BTU.
7. Per capita energy consumption is the amount of energy consumed per day.
8. Energy efficiency has a value between 0% and 100%.
9. Fossil fuels are preferred as raw energy sources because of their small energy densities.
10. "Sustainable Development" seeks to reach a balance between meeting current needs and preserving the ability of future generations to meet their needs.

Questions

1. What is the maximum value of the Human Development Index that a country can reach (the highest possible number)?
2. Which of the following companies was *not* one of the "Seven Sisters": Exxon, Royal Dutch/Shell, British Petroleum, Marathon, Texaco, Mobil, or Gulf Oil?
3. Name three non-fossil energy sources included in the global energy mix.
4. What are the three components of the United Nations Human Development Index?
5. Why did Britain switch to coal from wood as its primary energy source in the 16th century?

6. Use the following table to estimate the percent of world energy consumption that was due to fossil fuels in 2002.

Primary Energy Type	Total World Energy Consumption [Source: US EIA website, 2002]
Oil	39.9 %
Natural Gas	22.8 %
Coal	22.2 %
Hydroelectric	7.2 %
Nuclear	6.6 %
Geothermal, Solar, Wind & Wood	0.7 %

7. A measure of the energy that is available for doing useful work is called ________.
8. Which measure of energy is larger: one BTU or one megajoule?
9. What is Gross Domestic Product?
10. Why might a "have not" country with a large standing army be dangerous?

CHAPTER 2

FOSSIL ENERGY – COAL

Fossil fuels are the dominant energy source in the modern global economy. They include coal, oil, and natural gas. Coal is a combustible rock that is composed primarily of carbon-rich organic (carbonaceous) material. Coal contributed 22% to the mix of energy consumed by the United States in 2008 (Appendix B), and 27% to the mix of energy consumed in the world in 2006 (Appendix C). The formation and distribution of fossil fuels depends on the geologic history of the earth, which is where we begin our discussion of coal.

2.1 GEOLOGIC HISTORY OF THE EARTH

Many scientists believe that the earth first began to form 4 to 5 billion years ago. It is believed that the earth was a large body of hot, gaseous matter that whirled through space for several hundred million years. This body of matter slowly cooled and condensed. A solid crust gradually formed around the molten interior. The cross-section in Figure 2-1 bisects the earth and shows that the interior of the earth consists of an iron core wrapped inside a mantle of rock with a thin crust at the surface. Molten rock beneath the surface of the earth is called magma. If magma is expelled in a molten state by a volcano, it is called lava.

The thickness of the crust is small compared to the diameter of the earth. The crust consists of oceanic crust and continental crust (Figure 2-2). The mobile part of the upper mantle and crust is called the lithosphere. Lithospheric plates drift on a denser layer of semi-molten basalt called the asthenosphere.

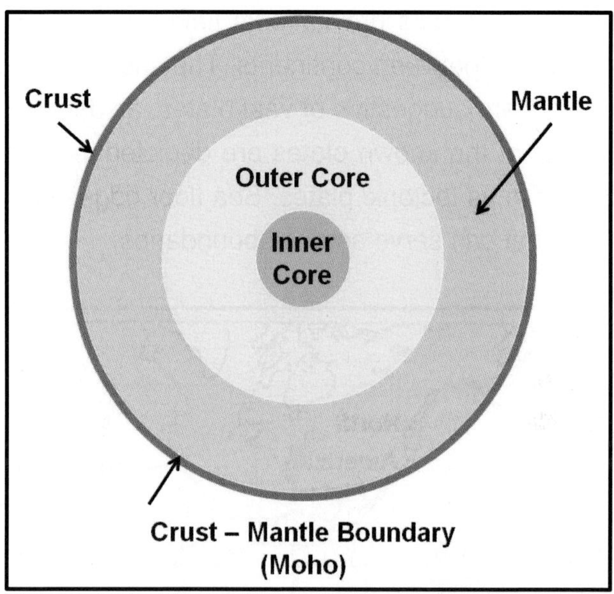

Figure 2-1. Schematic Cross-Section of the earth (not to scale)

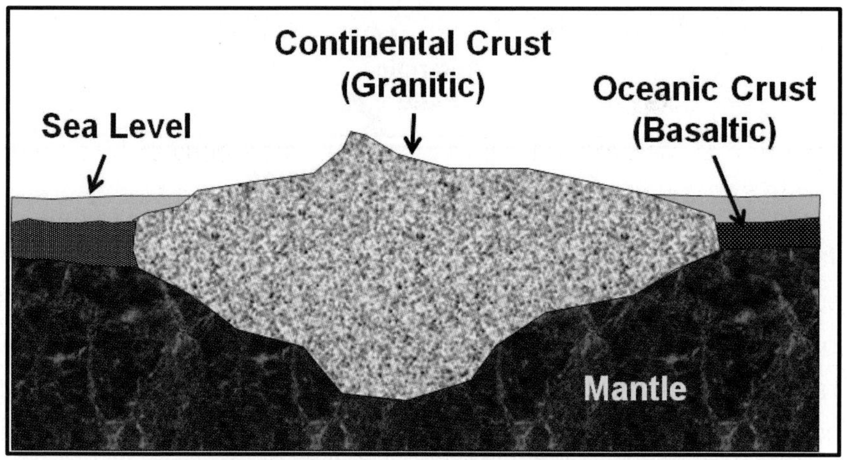

Figure 2-2. Cross-Section of the earth

As the earth cooled from its original state, the crust of the earth was subjected to forces that caused great changes in its topography, including the formation of continents and the uplift of mountain ranges. Pressure from the earth's interior could crack the sea floor and allow less dense molten material to flow onto the sea floor to form subsea ridges. Satellite measurements of the earth's gravitational field have identified sea floor ridges and boundaries between continents. The shapes of the boundaries between continents are suggestive of vast plates, as depicted in Figure 2-3. Only the largest of the known plates are depicted in the figure. These plates are referred to as tectonic plates. Sea floor ridges are long, subsea mountain ranges that can serve as plate boundaries.

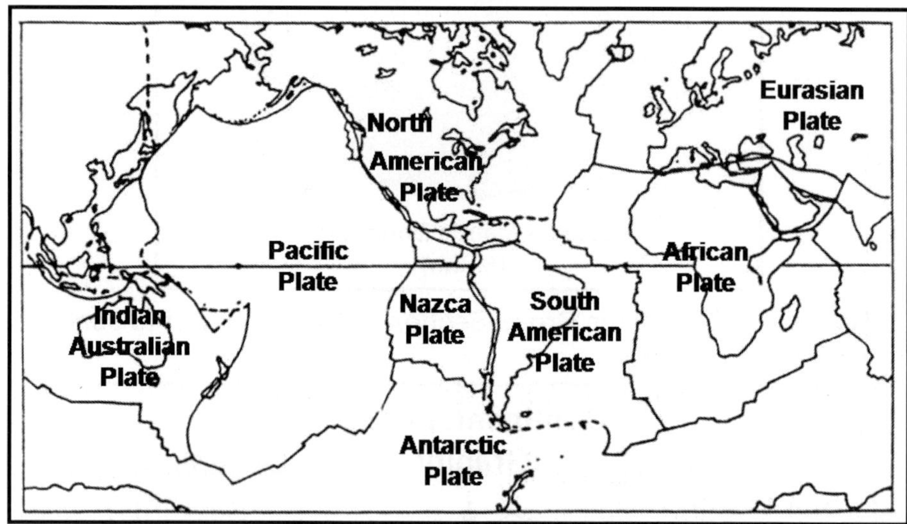

Figure 2-3. Tectonic Plates

In the theory of plate tectonics, the entire crust of the earth is considered a giant, ever-shifting jigsaw puzzle. Figure 2-4 shows the hypothesized movement of tectonic plates during the past 225 million years. It begins at the time that all surface land masses were thought to be coalesced into a single land mass known as Pangaea. Geoscientists believe that Pangaea was formed by the movement of tectonic plates, and the continued movement of plates led to the breakup of the single land mass into the surface features we see today. Tectonic plates move in relation to one another at the rate of up to 4 inches per year and are driven by forces which originate in the earth's interior. As the plates pull apart or collide, they can cause large-scale geologic events such as volcanic eruptions, earthquakes and mountain range formation.

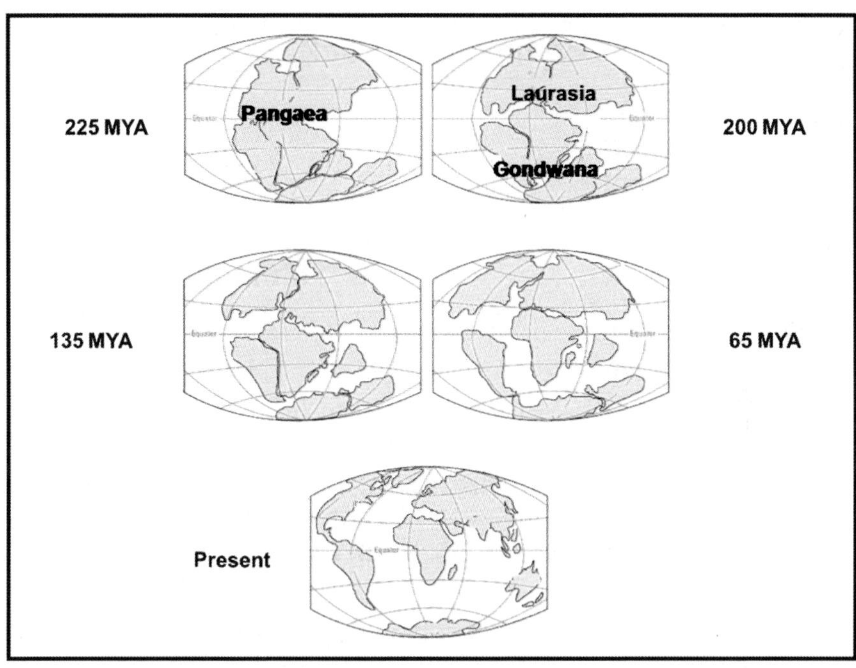

Figure 2-4. Tectonic Plate Movement [after USGS website, 2009]

The movement of tectonic plates across the surface of the globe generated forces powerful enough to cause rocks to form and change shape. Geologists recognize three primary types of rocks: igneous, sedimentary, and metamorphic. Igneous rocks are formed by the cooling and solidification of molten rock. Sedimentary rocks form when materials on the surface of the earth are weathered, transported, deposited, and cemented together. Sediment can be lithified, or made rock-like, by the movement of minerals into sedimentary pore spaces. A mineral is a naturally occurring, inorganic solid with a specific chemical and crystalline structure. In a process called cementation, minerals can form cement that binds grains of sediments together into a rock-like structure that has less porosity than the original sediment. Porosity is the fraction of void space between the grains in the material. Compaction occurs when pressure compresses the rock-like structure. Metamorphic rocks can form when buried sedimentary rocks are subjected to heat and pressure. This process of rock creation is known as the rock cycle, and is illustrated in Figure 2-5.

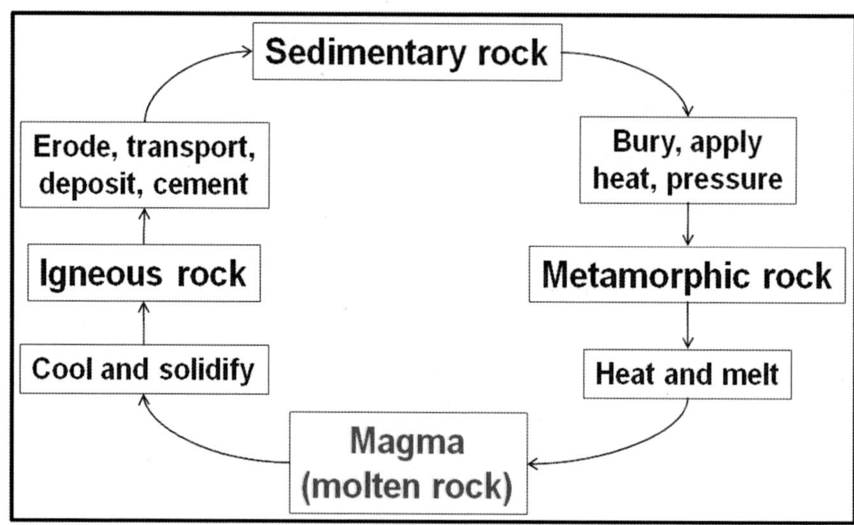

Figure 2-5. Rock Cycle

Slowly, over millions of years, tectonic plates move up and down relative to sea level, alternately causing erosion and deposition. Erosion occurs when chemical and physical processes cause exposed rock to break into smaller and smaller fragments. Wind and water transport these fragments from their source location. The fragments become finer as they collide with other objects during the transport process. A fragment will be deposited along with other fragments when the energy of the wind or water dissipates to the point where there is not enough energy to transport the fragment. The accumulation of fragments thickens as fragments are deposited in a specific location.

Deposition can range from thousands of feet of sediment in an area to none at all. Erosion can carve canyons, level once jagged mountains, or remove all traces of a formation that was once hundreds of feet thick. High pressure and temperature can cause rocks to change character in a process called metamorphism. Fragments may become fused together to form considerably larger objects. Given enough time, pressure and heat, rocks will melt and start the cycle again.

Typically, the formation of new rock occurs at plate boundaries, but it can also occur over "hot spots" within the earth's mantle. As plates collide, pressure and heat may cause part of the plate to melt, and result in molten rock being thrust to the surface. After cooling, surface rock is subjected to atmospheric phenomena.

Tectonic plate movement is considered the cause of extensive changes on the surface of the earth. Tectonic plate movement can lead to changes in sea level and the biosphere. Plate movement can lower sea level, creating a period of vast erosion and deposition. The biosphere is also affected by these changes. Plants and animals may thrive in one set of conditions, and readily become extinct when the conditions change. Plate movement also provides a mechanism to help explain the geographic distribution of organisms around the world. For example, the creation of a land bridge allows migration while the destruction of the land bridge can impose geographic isolation.

An ecosystem is composed of physical and biological components. Geologists use significant historical changes to the global ecosystem as markers on a geologic timeline. Table 2-1 shows an abridged version of the geologic time scale since the earth's formation. The geologic timeline is divided into several durations. Eons are subdivided into eras; eras are subdivided into periods; and periods are subdivided into epochs. The acronym MYBP in Table 2-1 stands for millions of years before the present. The range of starting times of selected intervals is reported from three references [Levin, 1991; Ridley, 1996; Press and Siever, 2001]. A comparison of the geologic time scales reported in the literature indicates that the actual chronology of the earth is approximate.

The oldest terrestrial rocks formed during the Precambrian Eon. Microscopic organisms are believed to have originated during this eon. Organisms with cellular nuclei (eukaryotes) appeared during the Proterozoic Era of the Precambrian.

The Cambrian Period in the Paleozoic Era of the Phanerozoic Eon is the era when life began to blossom. The demise of trilobites during the Permo-Triassic extinction marked the end of the Permian Period at the end of the Paleozoic Era. The Mesozoic Era includes the age of dinosaurs, and its end was marked by the Cretaceous-Tertiary (K-T) extinction. Cenozoic rocks are relatively new, being less than 70 million years old. Mammals began to flourish during the Cenozoic. The variations in life forms give rise to variations in the fossil record that can help geoscientists characterize the stratigraphic column, which shows the ordering of rock units as we drill down into the crust of the earth.

Table 2-1
Geologic Time Scale

Eon	Era	Period	Epoch	Approx. Start of Interval (MYBP)
Phanerozoic	Cenozoic	Quaternary	Recent (Holocene)	0.01
			Pleistocene	2
		Tertiary	Pliocene	5
			Miocene	24
			Oligocene	35 - 37
			Eocene	54 - 58
			Paleocene	65 - 66
	Mesozoic	Cretaceous		144
		Jurassic		206 - 213
		Triassic		245 - 251
	Paleozoic	Permian		286 - 300
		Carboniferous	Pennsylvanian	320
			Mississippian	354 - 360
		Devonian		408 - 409
		Silurian		438 - 439
		Ordovician		505 - 510
		Cambrian		543 - 590
Precambrian	Proterozoic			2500
	Archean			3800
	Hadean			

2.2 ORIGIN OF COAL

Coal was formed from organic debris by a process known as coalification. When some types of organic materials are heated and compressed over time, they can form water and gas and coal. In some cases, a high-molecular weight, waxy oil is also formed. For example, bog or swamp vegetation may be buried under anaerobic conditions and become peat. Peat is a coal precursor. It is an unconsolidated deposit of partially carbonized vegetable matter in a water-saturated environment. If peat is buried by rock in a depositional environment, it is subjected to increasing temperature and pressure. Volatile products and water migrate away from the formation. The resulting carbon-rich, compressed residue is coal. Figure 2-6 illustrates the coal formation process.

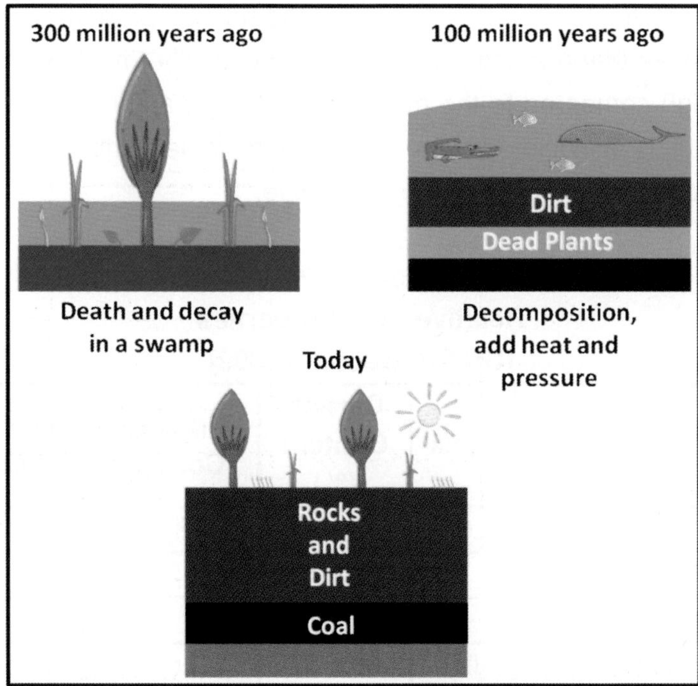

Figure 2-6. A Schematic View of Coal Formation [US EIA website, 2009]

Organisms that form coal when subjected to coalification include algae, phytoplankton, and zooplankton. Coal can also be formed by the bacterial decay of plants and, to a lesser extent, animals. Organic debris is composed primarily of carbon, hydrogen, and oxygen. It may also contain minor amounts of other elements such as nitrogen and sulfur. The organic origin of coal provides an explanation for the elemental composition of coal, which ranges from almost pure carbon to compounds of carbon with other elements, notably hydrogen, oxygen, and sulfur.

2.3 Coal

Coals are classified by rank. Rank is a measure of the degree of coalification or maturation of carbonaceous material. The lowest rank coal is lignite, followed in order by sub-bituminous coal, bituminous coal, anthracite and graphite. Table 2-2 summarizes the relative properties of coal for four coal ranks that are used as energy sources. The moisture content of lignite is high compared to the moisture content of anthracite. High moisture content is associated with low heating value, while low moisture content is associated with high heating value.

Table 2-2
Relative Coal Properties
[US EIA website, 2009]

Coal	Rank	Carbon Content (%)	Moisture Content	Heating Value
Lignite	Low	25 – 35	High	Low
Sub-bituminous		35 – 45		
Bituminous		45 – 86		
Anthracite	High	86 – 97	Low	High

Coal rank is correlated to the maturity, or age, of the coal. As a coal matures, the ratio of hydrogen to carbon atoms and the ratio of oxygen to carbon atoms decrease. The composition of the highest rank coal, graphite, approaches 100% carbon. Coal becomes darker and denser with increasing rank.

Coals burn better if they are relatively rich in hydrogen; this includes lower rank coals with higher hydrogen to carbon ratios. The percentage of volatile materials in the coal decreases as coal matures. Volatile materials include water, carbon dioxide and methane. Coal gas is gas absorbed in the coal structure or coal matrix. It is primarily methane with lesser amounts of carbon dioxide. The amount of gas that can be absorbed by the coal matrix depends on rank. As rank increases, the amount of methane in the coal matrix increases because the molecular structure of higher rank coals forms a structure that has a greater capacity to absorb gas and can therefore contain more gas.

Coals are organic, sedimentary rocks. Figure 2-7 shows an idealized representation of the physical structure of a coal seam. A coal seam is the stratum or bed of coal. It is a collection of coal matrix blocks bounded by natural fractures. The fracture network in coalbeds consists of microfractures called "cleats." An inter-connected network of cleats allows coal gas to flow from the coal matrix blocks when the pressure in the fracture declines. This is an important mechanism for producing coal gas.

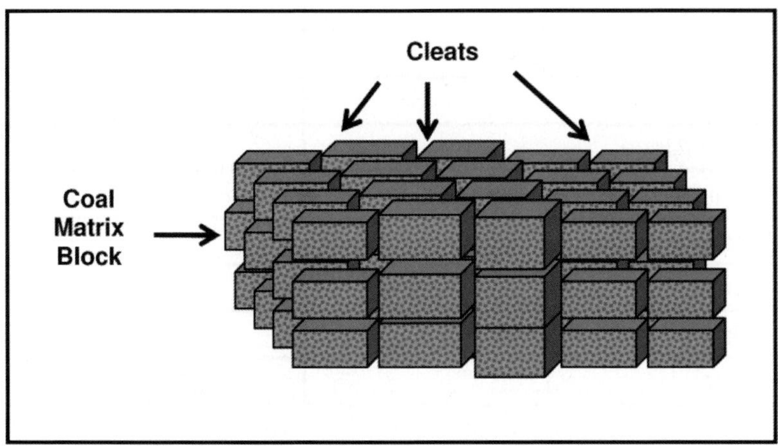

Figure 2-7. Schematic of a Typical Coal Seam

2.4 Distribution and Production of Coal

Coal is a non-renewable resource. The amount of coal is finite and takes more time to form than it does to consume. Consequently, the consumption of coal results in a reduction of the known supply. Since coal is a non-renewable resource, the amount of coal is expressed in terms of resources and reserves.

Coal resources include all coal that has been identified. Coal reserves refer to the subset of coal resources that is economically producible when they were classified. Improvements in technology and an increase in the price of coal can lead to the reclassification of coal from a resource to a reserve. The reserve concept applies to other non-renewable resources, including oil, gas and uranium.

Table 2-3
National Coal Reserves as Percent of Total Global Coal Reserves in 2005* [US EIA website, 2009]

Country	% of Total Global Coal Reserves
United States	28.4
Russia	18.6
China	13.6
Australia	9.1
India	6.7
South Africa	5.7
Ukraine	4.0
Kazakhstan	3.7
Former Serbia and Montenegro	1.6
Poland	0.9
Other	7.8
* Total global coal reserves in 2005 was approximately 930 billion short tons (1 short ton = 2000 lbs)	

Coal reserves for the ten countries with the largest reserves in 2005 are shown in Table 2-3. The ten countries represent over 92% of the coal reserves in the world. Reserves for all other countries are gathered in "Other."

Coal reserves for each country in Table 2-3 are expressed as a percent of total global coal reserves, which was approximately 930 billion short tons. A short ton is 2000 pounds or 907 kilograms. Many countries have negligible coal reserves. The United States, Russia, and China contain the largest percentage of global coal reserves.

Major coal regions in the United States are outlined in Figure 2-8. Lignite and anthracite are found predominantly in the Gulf Coast (Texas) and Northern Great Plains (North Dakota, South Dakota, and eastern Montana) regions. Sub-bituminous coal is located in the Rocky Mountain region. Bituminous coal is located in central Texas, central Colorado and Utah, the Interior region of Illinois, Missouri, and Iowa, and the Eastern region stretching from Alabama to Pennsylvania.

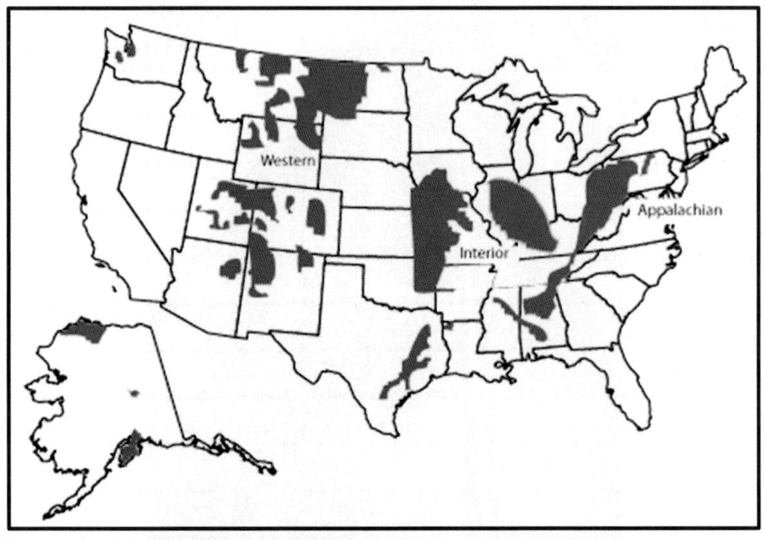

Figure 2-8. Major Coal Regions in the United States as of January 1, 2008 [US EIA website, 2009]

2.4.1 Coal Production

Coal is usually produced by extraction from coal beds. Mining is the most common extraction method. There are several types of mining techniques. Some of the more important coal mining techniques are strip mining, drift mining, deep mining, and longwall mining. Coal seams relatively close to the surface of the earth are strip mined. Consequently, strip mining is also known as surface mining. Any overlying material, such as topsoil and vegetation, is removed and coal is extracted by scraping. Drift mines are used to extract coal from coal seams that are exposed by the slope of a mountain. Drift mines typically have a horizontal tunnel entrance into the coal seam. Deep mining extracts coal from beneath the surface of the earth. In the case of deep mining, coal is extracted by mining the coal seam and leaving the bounding overlying layers and underlying layers undisturbed.

Figure 2-9. Coal Trains in Canada [Fanchi, 2002]

Coal is transported to consumers by ground transportation such as trains (Figure 2-9) and, to a lesser extent, ships. A relatively inexpensive means of transporting coal is the coal slurry pipeline. Coal slurry is a mixture of water and finely crushed coal. Coal slurry pipelines are not widely used because it is often difficult to obtain rights of way for coal slurry pipelines that extend over long distances, particularly in areas where a coal slurry pipeline would compete with an existing railroad right of way.

2.5 Fossil Energy and Combustion

Coal is a fossil fuel. The chemical energy in fossil fuels is released by the process of combustion. The complete combustion of fossil fuel with oxygen in the absence of nitrogen can be represented by the chemical reaction

$$\textit{Fuel} + \textit{Oxygen} \rightarrow \text{heat} + \text{carbon dioxide} + \text{water}$$

or

$$C_xH_y + \left(x + \frac{y}{4}\right) O_2 \rightarrow heat + xCO_2 + \frac{y}{2} H_2O$$

where x and y are stoichiometric coefficients. The stoichiometric coefficients for pure carbon are $x = 1,\ y = 0$ and the stoichiometric coefficients for methane are $x = 1,\ y = 4$.

Our air contains both nitrogen and oxygen. If a carbon-based fuel is burned in air, the carbon can react with oxygen and nitrogen to produce carbon dioxide, carbon monoxide, and nitrogen oxides. Nitrogen oxides produced by combustion of fossil fuels in air are abbreviated NOx, and include nitrogen-oxygen molecules with a single nitrogen atom, namely NO and NO_2. Water is also formed when hydrogen is present in the carbon-based fuel because hydrogen reacts with oxygen to form water.

Chemical reactions involved in fossil fuel combustion are exothermic reactions, that is, reactions that have a net release of energy. An exothermic reaction between two reactants *A*, *B* has the form

$$A + B \rightarrow \text{products} + \text{energy}$$

An exothermic reaction forms products with a net release of energy. The energy that must be added to a system to initiate a reaction is called activation energy.

Fossil fuels release a relatively large amount of energy during the combustion process, but they also release gases that are considered pollutants. The environmental impact of emission of combustion byproducts must also be considered.

Point to Ponder: Why is fossil fuel combustion a problem?
Fossil fuel combustion provided a new source of energy that helped prevent deforestation and now supports a relatively high quality of life in industrialized nations. Unfortunately, fossil fuel combustion also releases nitrogen oxides (NOx), carbon monoxide, and carbon dioxide.

NOx can react in the presence of sunlight with volatile organic compounds (VOCs) to form smog. VOCs are organic chemicals that can exist as gases in the atmosphere. Examples of VOCs include low molecular weight hydrocarbons such as methane and ethane, and formaldehyde. The smog can adversely affect a person's ability to breathe.

Carbon monoxide and carbon dioxide are examples of greenhouse gases. The presence of greenhouse gases in the atmosphere can trap heat energy. Many scientists believe that the additional heat is increasing the temperature of the atmosphere and is causing climate change. The environmental impact of fossil fuel combustion is discussed in more detail later.

2.6 Case Study: A Coal-Fired Power Plant

Coal-fired power plants generate electricity by burning coal to produce steam from water. Figure 2-10 illustrates the key systems of a coal-fired power plant. Flowing steam spins a turbine that turns a magnet in a generator to produce electricity. In a closed system, the steam is cooled, condenses into water, and the water is recycled to the boiler so that the process can be repeated. Generated electricity is transmitted through power lines to consumers. Electricity generation and distribution is discussed in more detail in Chapter 11.

An example of a coal-fired power plant is the Kingston Fossil Plant near Knoxville, Tennessee. The plant is managed by the Tennessee Valley Authority (TVA). According to the TVA [TVA, 2009], boilers at the plant heat water to about 1000°F (540°C) to create steam. The steam flows through pipes at pressures in excess of 1800 psi into turbines. The turbines spin at approximately 3600 revolutions per minute and drive generators that make alternating current electricity at a potential of 20,000 volts. River water provides cooling water to a condenser where steam exiting the turbines is cooled and condensed to water before it returns to the boiler.

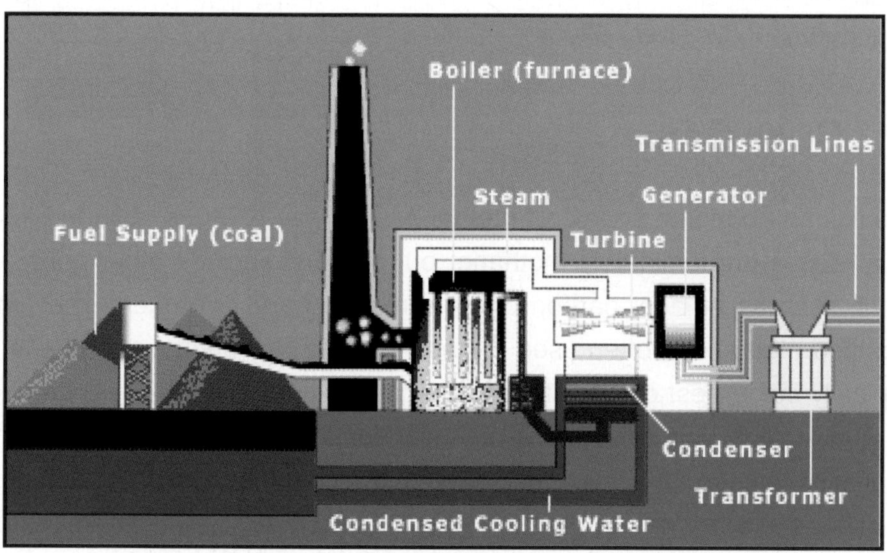

Figure 2-10. Tennessee Valley Authority [TVA, 2009]

Nine coal-fired boilers at the Kingston Fossil Plant burn a total of 5 million short tons of low-sulfur coal per year when operating at full capacity. It generates approximately 10 billion kilowatt-hours a year of electricity to supply energy to 670,000 homes. The net dependable generating capacity during the winter is 1450 megawatts. The amount of coal burned is approximately 14,000 short tons per day, which is enough to fill 140 railroad cars.

The TVA is investing US$6 billion dollars to control gas emissions from the plant. The emissions control equipment includes scrubbers to reduce sulfur dioxide (SO_2) emissions, and catalytic reduction systems to reduce NOx emissions.

In addition to gas emissions, coal-fired power plants produce ash, which is the residue left over from burning coal. A coal ash retention facility at the Kingston Fossil Plant spilled a large volume of sludge into the Emory River and surrounding countryside on December 22, 2008. According to knoxnews.com, the spill prompted several lawsuits for financial payments from the TVA to cover damages such as destroyed homes [see Barker, 2009a], a call by a state of Tennessee advisory board for tougher regulations of ash impoundments [Barker, 2009b], and a restructuring of management [Marcum, 2009].

2.7 Coal Gas

Coal is an abundant source of methane. Coal gas is a gas obtained from coal that is predominately methane, but can also include other constituents, such as ethane, carbon dioxide, nitrogen and hydrogen. The term coalbed methane is often used to describe coal gas because methane is typically the largest composition of the gas mixture. Coal gas is a product of coalification and exists as a monomolecular layer (a layer that is one molecule thick) on the internal surface of the coal matrix. Coal gas is able to diffuse into the natural fracture network when a pressure gradient exists between the matrix and the fracture network.

Gas recovery from coal begins with desorption of gas from the internal surface to the coal matrix and micropores. The gas then diffuses through the coal matrix and micropores into the cleats. Finally, gas flows through the cleats to the production well. The flow rate depends, in part, on the pressure gradient in the cleats and the density and distribution of cleats.

The controlling mechanisms for gas production from coal are the rate of desorption from the coal surface to the coal matrix, the rate of diffusion from the coal matrix to the cleats, and the rate of flow of gas through the cleats.

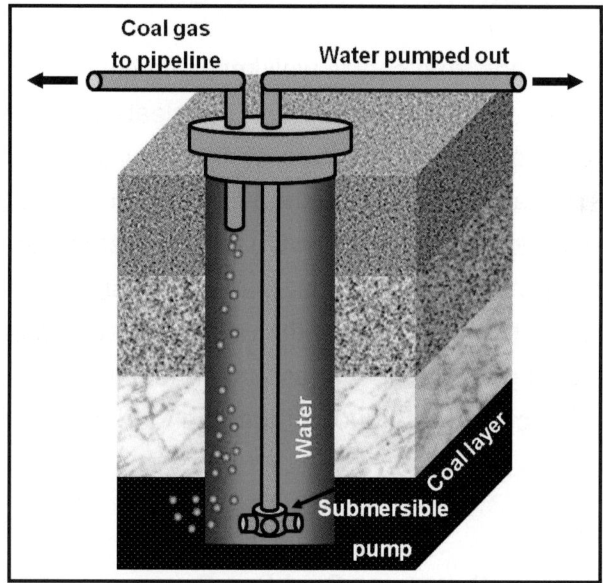

Figure 2-11. Schematic of Coal Gas Well

Figure 2-11 is a schematic of a coal gas well. Coal gas production often requires the production of large volumes of water before significant volumes of coal gas will flow. Additional energy may be needed to lift the water to the surface. This additional energy is provided by a submersible pump in Figure 2-11.

Coal gas recovery can be increased by injecting carbon dioxide (CO_2) into the coal seam. Carbon dioxide is generated in large quantities by a variety of processes. For example, CO_2 is produced by biological, industrial and transportation processes. In particular, we exhale CO_2 with every breath, and CO_2 is a by-product of combustion in power plants. Coal absorbs more CO_2 than methane so the injection of CO_2 into a coalbed can accomplish two objectives: increase methane production from coal, and store CO_2 in coal. Storage of CO_2 in coal reduces the amount of CO_2 that might otherwise be emitted into the atmosphere. The presence of CO_2 in the atmosphere can have an adverse environmental impact that is discussed in more detail in later.

2.7.1 Global Distribution of Coal Gas

The out-gassing of gas from coal is well known to coal miners as a safety hazard, and occurs when the pressure in the cleat system declines. The gas in the microscopic pore structure of the coal is now considered a source of natural gas. Coal gas is found in the coal seams presented in Table 2-3. The World Coal Institute (WCI) reports that the largest coal gas resource bases are in Russia, China, Canada, Australia and the United States. Table 2-4 lists an estimate of the coal gas resource base for these countries. The total global coal gas resource base in 2006 was estimated to be 5 quadrillion cubic feet. Much of the coal gas resource is underdeveloped.

Table 2-4
Nations with Largest Coal Gas Reserves in 2006 [WCI, 2009]

Country	Estimated Coal Gas Resource Base (trillion cubic feet)
Canada	600 to 3200
Russia	600 to 2800
China	1000 to 1200
Australia	280 to 490
United States	140 to 390

2.8 ACTIVITIES

True-False

Specify if each of the following statements is True (T) or False (F).

1. An exothermic reaction releases energy.

2. Porosity is the fraction of void space in rock.
3. An endothermic chemical reaction forms products with a net release of energy.
4. Molten rock beneath the surface of the earth is called lava.
5. Molten rock cools and solidifies into igneous rock as part of the rock cycle.
6. Tectonic plate movement can affect sea levels, rock creation, and the biosphere.
7. The dinosaurs became extinct during the Permo-Triassic extinction.
8. Coals are organic, metamorphic rocks.
9. Strip mining, drift mining, deep mining and longwall mining are common types of coal mining.
10. Coal gas is primarily methane.

Questions

1. How are coals classified?
2. What is the primary negative byproduct of combustion?
3. What is a cleat in a coal seam?
4. What hydrocarbon gas can be found in coalbeds in abundant amounts?
5. What are the three primary rock types?
6. List the four stages of the rock cycle.
7. What is coalification?
8. What is the difference between a coal resource and a coal reserve?
9. How do coal-fired power plants generate electricity?
10. How is CO_2 used in coal gas production?

CHAPTER 3

FOSSIL ENERGY – OIL AND GAS

Oil and gas are terms that refer to mixtures of hydrocarbon molecules in the liquid phase and gas phase, respectively. Table 3-1 lists the most common elements found in crude oil and natural gas. Crude oil is "a mixture of hydrocarbons that exists in liquid phase in natural underground reservoirs and remains liquid at atmospheric pressure" after passing through facilities on the surface that separate gas and liquid [US EIA website, 2009].

Table 3-1
Elemental Composition of Crude Oil and Natural Gas

Element	Composition (% by mass)
Carbon	84% - 87%
Hydrogen	11% - 14%
Sulfur	0.6% - 8%
Nitrogen	0.02% - 1.7%
Oxygen	0.08% - 1.8%
Metals	0% - 0.14%

The actual elemental composition depends on factors such as molecular composition of the source, reservoir temperature and reservoir pressure. Hydrocarbon molecules in oil and gas are organic molecules. Non-hydrocarbon molecules include nitrogen, carbon dioxide, and hydrogen sulfide.

Another term for hydrocarbon mixtures is petroleum. Petroleum liquid typically refers to crude oil and natural gas plant liquids. Natural gas is typically methane with lesser amounts of heavier hydrocarbon molecules like ethane and propane. If the natural gas contains liquids dissolved in the gas phase at high temperatures and pressures, the liquids can condense out of the gas phase when the temperature and pressure of the gas is reduced. The change in temperature and pressure typically occurs when gas is produced from a subsurface formation to a facility on the surface. The liquids obtained from the gas are referred to as condensates or natural gas plant liquids.

Large segments of modern economies are designed to use oil and gas, even though oil and gas are considered finite, non-renewable resources. The mix of energy consumed by the United States in 2008 included over 44% petroleum liquid and over 28% natural gas (Appendix B). The mix of energy consumed in the world in 2006 included over 36% petroleum liquid and over 23% natural gas (Appendix C). Most crude oil is refined for use in transportation, while gas is used to generate electric power and as a fuel in industrial, commercial, and residential sectors of the economy.

We consider conventional and unconventional oil and gas resources in this chapter. The United States Energy Information Administration defines conventional oil and gas resources as "crude oil and natural gas that is produced by a well drilled into a geologic formation in which the reservoir and fluid characteristics permit the oil and natural gas to readily flow to the wellbore" [US EIA website, 2009]. Unconventional oil refers to hydrocarbon production from low permeability shale (shale oil) and tar sands. Unconventional gas refers to gas production from coal (coal gas), low permeability sands (tight gas), and low permeability shale (shale gas). The primary difference between conventional and unconventional oil and gas is the ability of the fluid to flow through rock.

The objective of this chapter is to describe how oil and gas resources are formed and distributed throughout the world. We begin by discussing the geologic environment.

Point to Ponder: What is petroleum?
The term petroleum is widely used but inconsistently defined. The Society of Petroleum Engineers (SPE) defines petroleum as "naturally occurring liquids and gases which are predominately comprised of hydrocarbon compounds" [SPE Definitions, 2009]. By contrast, the United States Energy Information Administration defines petroleum as "a broadly defined class of liquid hydrocarbon mixtures. Included are crude oil, lease condensate, unfinished oils, refined products obtained from the processing of crude oil, and natural gas plant liquids." [US EIA website, 2009]. Both SPE and EIA define petroleum as naturally occurring mixtures of hydrocarbon molecules. The SPE defines petroleum in terms of liquids and gases, while EIA defines petroleum in terms of liquids and keeps gas separate. It is important to recognize context to understand the meaning of the term petroleum.

3.1 Geology of Oil and Gas Reservoirs

Naturally occurring hydrocarbon mixtures are usually found in the pore space of sedimentary rocks. According to the rock cycle, sedimentary rocks are formed by erosion, transport, deposition, and cementing of igneous rock. Unlike sedimentary rocks, igneous and metamorphic rocks originated under conditions of high pressure and temperature that did not favor the formation or retention of petroleum. Any oil and gas that might have occupied the pores of a metamorphic or igneous rock would ordinarily be cooked away by heat and pressure. Consequently, we focus our attention on sedimentary rock formations as geologic environments for containing oil and gas.

3.1.1 Rock Attributes

The key attributes used to classify sedimentary rock are mineral composition, grain size, color, and structure. Each of these attributes is considered here.

Mineral content is an important characteristic of a rock. The minerals that comprise a rock help determine the chemical and physical properties of the rock. For example, quartz may be able to withstand multiple cycles of erosion and deposition since quartz is much less reactive than other minerals. As another example, the presence of clays in the pore space can cause reduced productivity if a change in the salinity of the formation causes clay swelling. Fluids introduced into the reservoir through the wellbore can react with clays to swell and plug clay-bearing formations.

Grains which form sedimentary rocks are created by weathering processes at the surface of the earth. Weathering creates fragments that can be practically any size, shape, or composition. A glacier may create and transport a fragment the size of a house, while a desert wind might create a uniform bed of very fine sand. The fragments, also known as sediments, are transported to the site of deposition. Mineral transport is often achieved by water-based processes such as rivers, floods, and tides. When fragments are transported over long distances, only the most durable fragments survive the transport. Grains which started out as angular chunks of rock slowly become smaller and more rounded as they roll and bump along the transport pathway.

The shape of rock grain edges tells us something about the history of a rock. If rock edges are sharp, the rock probably did not get transported very far. Rounded edges indicate a longer period of transport. Rocks made up of rounded edges may have better permeability than rocks that are flat or have sharp edges.

Sorting refers to the uniformity of grain size. Fluids typically flow better through a well-sorted rock than a poorly-sorted rock. The ability to flow is related to a rock property known as permeability. Fluid flows faster through a rock with large permeability than it does through a rock with low permeability. Permeability is an important property that controls flow, and has a direct impact on the economic viability of producing fluids from a formation.

3.1.2 Depositional Environment

The environment under which a rock forms is called the environment of deposition. As the environment moves from one location to another, it leaves a laterally continuous progression of rock which is distinctive in character. For example, the rise or fall of sea level relative to a continental plate can change shorelines and their associated environments of deposition. If the progression of rock is large enough to be mapped, it can be called a formation.

Formations are the basic descriptive unit for a sequence of sediments. The formation represents a recognizable, mappable rock unit that was deposited under a uniform set of conditions at one time. Formations can extend for miles and vary in thickness from a few feet to hundreds of feet. The thickness is related to the length of time an environment was in a particular location, and how much subsidence was occurring during that period. An example of a stratified rock column is shown in Figure 3-1.

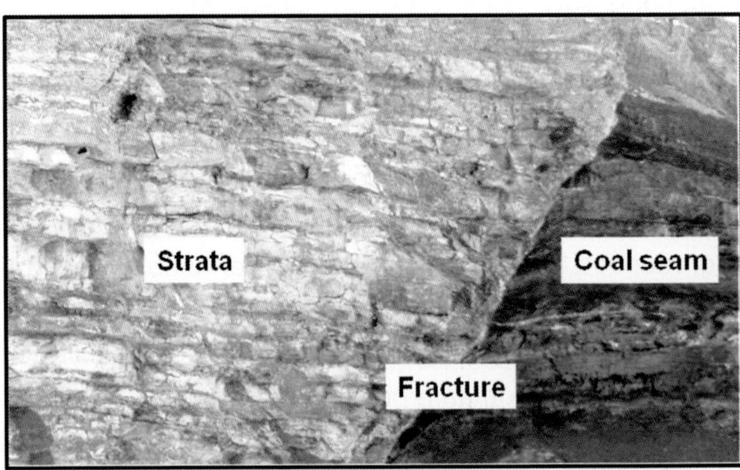

Figure 3-1. Example of a Stratified Rock Column

Each environment of deposition causes a different type of rock sequence to be deposited. The Mississippi River is an example of a fluvial (river) environment of deposition, and the mouth of the Mississippi River is a deltaic environment. Sandstone can be deposited in both a fluvial environment and a deltaic environment. The deposited sandstone is different in each environment. A fluvial system can produce rocks that meander along, while a delta tends to stay in one place to enable a large volume of sediment to be deposited. In a fluvial system, the grain size of deposited sand is finer at shallower depths. In addition, there is a coarser deposit at the base of the formation that distinguishes the fluvial system from a delta deposit. Lying above the sandstones of a meandering river are typically mudstones that may have desiccation cracks and traces of roots. The sandstones are interpreted to be formed in the river channel, while the mudstones are considered floodplain deposits.

The term facies refers to those characteristics of rocks that can be used to identify their depositional environment. In the fluvial deposition example, sandstone and mudstone are distinct facies.

3.2 Origin of Oil and Gas

Sedimentary rock formations are the most common geologic environments for storing oil and gas, but that does not mean that sedimentary rock formations are the geologic environments where oil and gas originated. Two theories of the origin of fossil fuels are considered here: the biogenic theory, and the abiogenic theory.

3.2.1 Biogenic Theory

The biogenic theory is the mainstream scientific view of the origin of fossil fuels. In the biogenic theory, fossil fuels are formed by a type of biochemical precipitation called organic sedimentation. We saw in Chapter 2 that a carbon rich organic material called peat can be formed when vegetation dies and decays in aqueous environments such as swamps. If peat is buried by subsequent geological activity, the buried peat is subjected to increasing temperature and pressure, and peat can eventually be transformed into coal by the process of diagenesis. Diagenesis denotes the physical and chemical changes that occur during the conversion of sediments to rock. A similar diagenetic process is thought to apply to the origin of oil and gas.

Hydrocarbon mixtures can exist in solid, liquid or gas phase in the reservoir. The phase depends on the composition of the mixture and the temperature and pressure of the reservoir. The elemental mass content of naturally occurring hydrocarbon mixtures ranges from approximately 84% to 87% carbon and 11% to 14% hydrogen, which is comparable to the carbon and hydrogen content of life. This is one piece of evidence for the origin of petroleum from biological sources.

The biochemical process for the formation of oil and gas is illustrated in Figure 3-2. It begins with the death of microscopic organisms such as algae and bacteria. The remains of the organisms settle into the sediments at the base of an aqueous environment as organic debris. Lakebeds and seabeds are examples of favorable sedimentary environments. Subsequent sedimentation buries the organic debris.

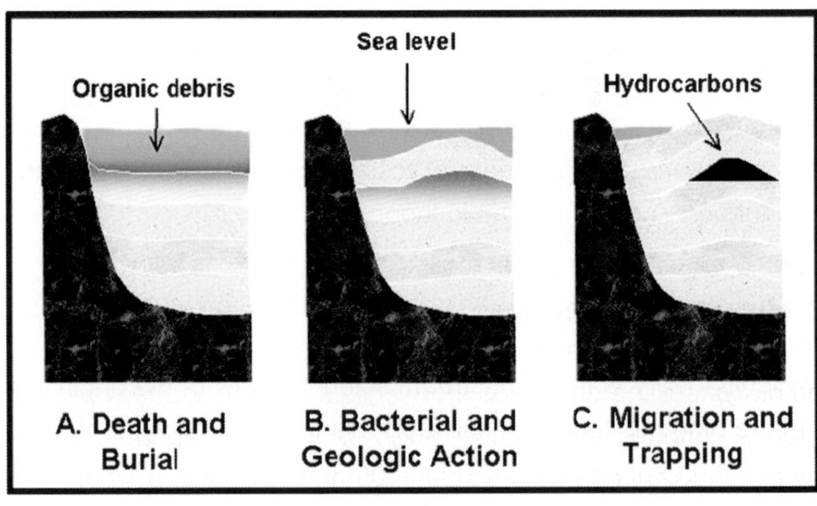

Figure 3-2. Biogenic Origin of Oil and Gas

The rise and fall of sea level, as well as other geologic processes, continue the process of burying the organic debris. As burial continues, the organic material is subjected to increasing temperature and pressure, and is transformed by bacterial action into oil and gas. Petroleum fluids are usually less dense than water and will migrate upwards until they encounter impermeable barriers and are collected in geologic traps. The accumulation of a hydrocarbon mixture in a geologic trap forms an oil and gas reservoir.

3.2.2 Abiogenic Theory

In the biogenic theory, the origin of oil and gas begins with the death of organisms that live on or near the surface of the earth. An alternative hypothesis called the abiogenic theory says that processes deep inside the earth, in the earth's mantle, form petroleum fluids. Thomas Gold, an advocate for the abiogenic theory, argued that simple inorganic and organic molecules in the interior of the forming earth were subjected to increasing heat and pressure, and eventually formed more complex molecules.

Gold [1999] presents several pieces of evidence in support of the abiogenic theory, and the possible existence of a biological community deep inside the earth. Some of Gold's evidence includes the existence of microbial populations that can thrive in extreme heat. These microbes, notably bacteria and archaea, grow at hot, deep ocean vents and can feed on hydrogen, hydrogen sulfide, and methane. Gold considers life forms at deep ocean vents transitional life forms that exist at the interface between two biospheres.

One biosphere is the surface biosphere and includes life that lives on the continents and in the seas on the crust of the earth. Gold postulates that a second biosphere exists in the mantle of the earth. He calls the second biosphere the deep biosphere. The surface biosphere uses chemical energy extracted from solar energy, while the deep biosphere feeds directly on chemical energy. Oxygen is a requirement in both biospheres. Gold's deep biosphere is the source of life that eventually forms hydrocarbon mixtures in the earth's mantle. Crustal oil and gas reservoirs are formed by the upward migration of petroleum fluid until the fluid is stopped by impermeable barriers and accumulates in geological traps. The abiogenic theory is illustrated in Figure 3-3.

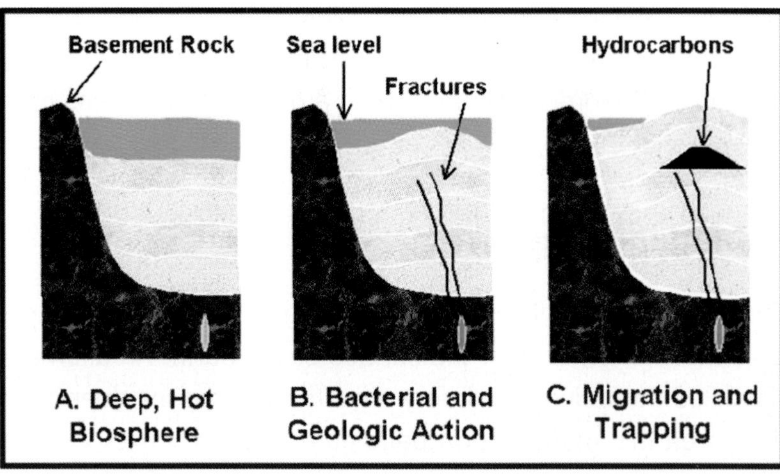

Figure 3-3. Abiogenic Origin of Oil and Gas

Point to Ponder: Why does it matter whether the biogenic theory or the abiogenic theory is right?
Two of the arguments driving a transition from fossil energy to other forms of energy are the belief that the earth contains a finite amount of fossil fuels, and that fossil fuels are not produced by natural processes fast enough to allow fossil fuels to be used as an inexhaustible source of energy. These arguments are based on the assumption that the biogenic theory is correct. If the abiogenic theory is correct, existing estimates of the amount of petroleum fluid and the rate at which it is renewed could be significantly understated. There is still an environmental argument for reducing our dependence on fossil fuels that is known as climate change. It is considered in more detail later.

3.3 Oil and Gas Reservoirs and Reserves

Several key factors must be present to allow the development of a hydrocarbon reservoir:

1. A source for the hydrocarbon must be present. For example, the biogenic theory says that oil and gas are formed by the decay of single-

celled aquatic life. Shales formed by the heating and compression of silts and clays are often good source rocks. Oil and gas can form when the remains of an organism are subjected to increasing pressure and temperature.

2. A flow path must exist between the source rock and reservoir rock.
3. Once hydrocarbon fluid has migrated to a suitable reservoir rock, a trapping mechanism must exist. If the hydrocarbon fluid is not stopped from migrating, buoyancy and other forces will cause it to move towards the surface.
4. Overriding all of these factors is timing. A source rock can provide large volumes of oil or gas to a reservoir, but a trap must exist at the time oil or gas enters the reservoir so that fluid can accumulate.

If all of these conditions are met, a hydrocarbon bearing reservoir can form.

3.3.1 Resource Pyramid

A discovery well is the first well drilled into an exploration prospect that finds hydrocarbons. After the discovery well, the reservoir becomes a resource that may or may not be economic. The value of a resource can be displayed using the resource pyramid (Figure 3-4).

A resource container (such as a coal seam, oil reservoir, gas reservoir, uranium ore, gold ore, etc.) is placed at the top of the pyramid if it contains a relatively large concentration of resource that is relatively inexpensive to extract. By contrast, a resource container is placed lower in the pyramid if the concentration of the resource decreases or the extraction becomes more difficult or expensive. If the resource container has a relatively small concentration of resource and is difficult to extract, it will be placed near the base of the pyramid. As an illustration, a large, shallow, light oil reservoir in an easily accessible part of the earth would be at the top of the pyramid, while a small, deep, heavy oil reservoir in an extreme environment such as the Arctic would be at the base of the pyramid. A pyramid is used because the amount of resource that is both easy to extract and has a high concentration is expected to be small, while the amount of resource that is found in a low concentration container and is difficult to extract is expected to be large.

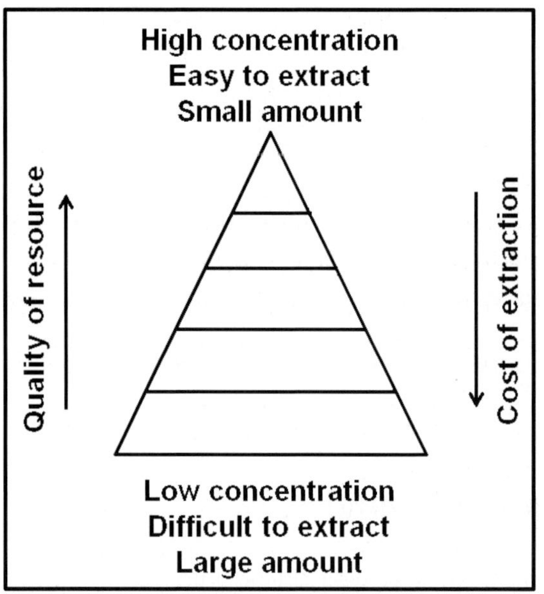

Figure 3-4. Illustration of a Resource Pyramid

Today, many people in the energy industry say that the easy oil has been found. This comment is an indirect reference to the resource pyramid for oil. In general, resources near the apex of the resource pyramid are considered easy, while those near the base of the resource pyramid are considered difficult.

3.3.2 Reserves

The amount of fluids in hydrocarbon bearing reservoirs is known as reserves. The definition of reserves adopted by two influential organizations is summarized in Table 3-2 [SPE-PRMS, 2007]. SPE is the Society of Petroleum Engineers, WPC is the World Petroleum Congress, and PRMS is the Petroleum Resources Management System. Reserve estimates have many commercial applications. For example, reserve estimates are used to compare projects, establish the value of a resource, and support applications for project financing.

Table 3-2
SPE/WPC Reserves Definitions

Proved Reserves	• Those quantities of petroleum, which by analysis of geoscience and engineering data, can be estimated with reasonable certainty to be commercially recoverable, from a given date forward, from known reservoirs and under defined economic conditions, operating methods, and government regulations. • There should be at least a 90% probability (P_{90}) that the quantities actually recovered will equal or exceed the estimate.
Probable Reserves	• Those additional reserves which analysis of geoscience and engineering data indicate are less likely to be recovered than Proved Reserves but more certain to be recovered than Possible Reserves. • There should be at least a 50% probability (P_{50}) that the quantities actually recovered will equal or exceed the estimate.
Possible Reserves	• Those additional reserves which analysis of geoscience and engineering data suggests are less likely to be recoverable than Probable Reserves. • There should be at least a 10% probability (P_{10}) that the quantities actually recovered will equal or exceed the estimate.

Oil and gas reserves are not directly measured. They must be estimated from a limited set of observations. The resulting value of reserves is an estimate that should include our lack of knowledge, or uncertainty, about important factors such as size of resource. The uncertainty may be expressed in terms of a probability of actually attaining the estimate of reserves. We illustrate this concept by associating a normal probability distribution with the SPE-PRMS reserves definitions. For a normal distribution with a mean and standard deviation (abbreviated std dev), the SPE-PRMS reserves definitions are

$$\text{Proved Reserves} = P_{90} = \text{mean} - 1.28\times(\text{std dev})$$

$$\text{Probable Reserves} = P_{50} = \text{mean}$$

$$\text{Possible Reserves} = P_{10} = \text{mean} + 1.28\times(\text{std dev})$$

An estimate of reserves is needed to compare the commercial values of different reservoirs. Uncertainty in this case is captured by the value of the standard deviation. A large standard deviation is associated with a large uncertainty in reserves estimation.

Figure 3-5 shows a normal distribution for a mean of 200 million barrels of oil (MMSTBO) and three different standard deviations (20 MMSTBO, 40 MMSTBO, and 60 MMSTBO). The difference between possible and proved reserves gets smaller as the standard deviation decreases.

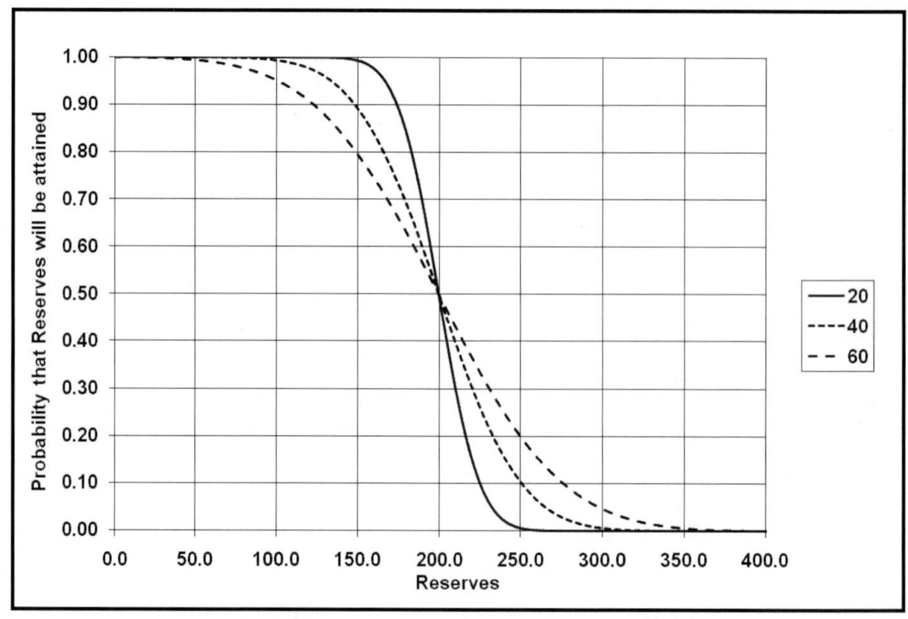

Figure 3-5. Illustration of a Reserves Distribution

3.4 CLASSIFICATION OF OIL AND GAS

Hydrocarbon fluids that are one phase at reservoir temperature and pressure often become two-phase fluids when they are produced to surface temperature and pressure. Natural gas is a hydrocarbon mixture in the gaseous state at surface conditions. Crude oil is a hydrocarbon mixture in the liquid state at surface conditions. Heavy oils do not contain much gas in solution at reservoir conditions and have a relatively large molecular weight. By contrast, light oils typically contain a large amount of gas in solution at reservoir conditions and have a relatively small molecular weight.

A classification of hydrocarbon fluid types is given in Table 3-3. Separator gas-oil ratio (GOR) is a useful indicator of fluid type. It is the ratio of gas produced at the surface to the liquid produced. The unit SCF/STB equals one standard cubic foot (SCF) of gas per stock tank barrel (STB) of oil. The gas and oil volumes expressed in SCF and STB are at surface temperature and pressure.

Table 3-3
Rules of Thumb for Classifying Hydrocarbon Fluid Types

Fluid Type	Separator GOR (SCF/STB)	Behavior when Pressure Drops
Dry gas	No surface liquids	Remains gas
Wet gas	> 100,000	Remains gas
Condensate	3,000 – 100,000	Gas with liquid drop out
Volatile oil	1,500 – 3,000	Liquid with significant gas
Black oil	100 – 1,500	Liquid with some gas
Heavy oil	≈ 0	Negligible gas formation

Another way to classify hydrocarbon liquids is to compare the properties of the hydrocarbon liquid to water. Two key properties are viscosity and density. Viscosity is a measure of the ability to flow, and density is the amount of material in a given volume. Water viscosity is 1 cp (centipoise) and water density is 1 gram per cubic centimeter (1 g/cc) at 60°F. A liquid with smaller viscosity than water flows more easily than water. Gas viscosity is much less than water viscosity. Tar, on the other hand, is a very high viscosity relative to water.

A comparison of hydrocarbon liquid density with water density is made using API gravity, where API refers to the American Petroleum Institute. The ratio of hydrocarbon liquid density (HC liq density) to water density (water density) is the specific gravity (spec grav) of the hydrocarbon liquid. Specific gravity of a hydrocarbon liquid is calculated as

$$\text{HC liq density} = (\text{water density})/(\text{spec grav})$$

API gravity is calculated using specific gravity:

$$\text{API} = (141.5/\text{spec grav}) - 131.5$$

The API gravity of water is 10°API (called 10 degrees API).

Table 3-4
Classifying Hydrocarbon Liquid Types Using API Gravity and Viscosity

Liquid Type	API Gravity (degrees API)	Viscosity (centipoises)
Light oil	> 31.1	
Medium oil	22.3 to 31.1	
Heavy oil	10 to 22.3	
Water	10	1 cp
Extra heavy oil	4 to 10	< 10,000 cp
Bitumen	4 to 10	> 10,000 cp

Table 3-4 shows a hydrocarbon liquid classification scheme using API gravity and viscosity. Water properties are included in the table for comparison. Bitumen is a hydrocarbon mixture with large molecules and high viscosity. Crude oil usually refers to light oil, medium oil, and heavy oil.

Crude oil is less dense than water, while extra heavy oil and bitumen are denser than water. In general, crude oil will float on water, while extra heavy oil and bitumen will sink in water.

3.5 Shale Oil, Tar Sands and Extra Heavy Oil

Shale oil is solid bituminous material contained in low permeability shale. An oily liquid is obtained when the material is heated. Tar sands, also known as natural bitumen, are a combination of clay, sand, water, and bitumen. The World Energy Council [WEC, 2007] estimated that the shale oil resource base in mid-2005 was approximately 2.8 trillion barrels. Nations with the largest volume of shale oil are listed in Table 3-5.

Table 3-5
Nations with Largest Shale Oil Resources in mid-2005 [WEC, 2007, Table 3-1]

Shale Oil Resource (billion barrels)	
United States	2085
Russia	248
Congo	100
Brazil	82
Italy	73

The World Energy Council [WEC, 2007] estimated that the total original oil in place of natural bitumen (tar sand) and extra heavy oil in mid-2005 were approximately 3.3 trillion barrels and 2.5 trillion barrels, respectively. Nations with the largest volumes of natural bitumen and extra heavy oil in mid-2005 are listed in Table 3-6.

Total original oil in place listed in Table 3-6 is an estimate of the size of the resource. It does not represent the amount of resource that will be developed. We can use an estimate of resource size as one factor in prioritizing the importance of the resource.

Table 3-6
Nations with Largest Volumes of Natural Bitumen and Extra Heavy Oil in mid-2005
[WEC, 2007, Tables 4-1 and 4-2]

Natural Bitumen (Tar Sands) Total Original Oil in Place (billion barrels)		Extra Heavy Oil Total Original Oil in Place (billion barrels)	
Canada	2397	Venezuela	2446
Kazakhstan	421	United Kingdom	11.8
Russia	347	China	8.8
United States	53	Azerbaijan	8.9
Nigeria	38	Italy	2.7

Although difficult to produce, the amount of low API gravity hydrocarbon represented in Tables 3-5 and 3-6 has motivated efforts to improve production techniques. Shale oil, tar sands and extra heavy oil can be extracted by mining when they are close enough to the surface. Tar pits have been found around the world and have been the source of many fossilized dinosaur bones. In locations where oil shale, tar sands, and extra heavy oil are too deep to mine, it is necessary to increase the mobility of the hydrocarbon.

An increase in permeability or a decrease in viscosity can increase fluid mobility. Increasing the temperature of shale oil, tar or asphalt can significantly reduce viscosity. If there is enough permeability to allow injection, steam or hot water can be used to increase formation temperature and reduce hydrocarbon viscosity. Steam Assisted Gravity Drainage (SAGD) is used to increase recovery of oil from tar sands in Canada. In many cases, however, permeability is too low to allow significant injection of a heated fluid. An alternative to fluid injection is electromagnetic heating. Radio frequency heating has been used in Canada, and electromagnetic heating techniques are being developed for other parts of the world.

3.6 UNCONVENTIONAL GAS

Environmental concerns are motivating a change from fossil fuels to an energy supply that is clean. Clean energy refers to energy that has little or no detrimental impact on the environment. Natural gas is a source of relatively clean energy that can be obtained from a variety of sources including conventional oil and gas fields, unconventional gas resources, landfill gas, and municipal solid waste gas (MSW gas). Unconventional gas resources include gas hydrates, tight gas sands, coal gas, and shale gas. Jenkins and Boyer [2008] reported that production of coal gas and shale gas in the United States was approximately 2.7 trillion standard cubic feet of gas, or about 15% of total net gas production. Approximately 1.7 trillion standard cubic feet of gas comes from over 40,000 coalbed wells, and 1.0 trillion standard cubic feet of gas comes from over 40,000 shale gas wells. Coal gas was discussed in Chapter 2. Gas hydrates, tight gas sands and shale gas are discussed below. Gas from landfills and municipal solid waste (MSW) gas are obtained from the decay of organic waste. Landfill gas and MSW gas are not fossil fuels; they are renewable energy sources.

3.6.1 Gas Hydrates

Gas hydrates have traditionally been considered a problem for oil and gas field operations, but their potential commercial value as a clean energy resource is changing industry perception. Gas hydrates are naturally occurring, crystalline complexes that are formed when one type of molecule completely encloses another type of molecule in a lattice. In the case of gas hydrates, hydrogen-bonded water molecules form a cage-like structure in which molecules of gas (such as methane, ethane and propane) are absorbed or bound. Figure 3-6 illustrates methane hydrate with a methane molecule in the center of the lattice.

Methane hydrates contain a relatively large volume of methane in the hydrate complex. The hydrates complex contains about 85 mole percent water and approximately 15 mole percent guests, where a guest molecule is methane or some other relatively low molecular weight hydrocarbon.

Methane hydrates are found around the world. They exist on land in sub-Arctic sediments and on seabeds where the water is near freezing. They can be found in Arctic sands, marine sands, fractured muds, and shales. Recognizing that the estimate of global hydrate resource size is very uncertain, Boswell [2009] reported that gas hydrates may contain on the order of 680,000 trillion cubic feet of methane. For comparison, natural gas consumption in the United States is approximately 23 trillion cubic feet per year. The size of the gas hydrate resource is a huge target for future development. Difficulties in cost-effective production of methane hydrates have hampered the production of methane from hydrates.

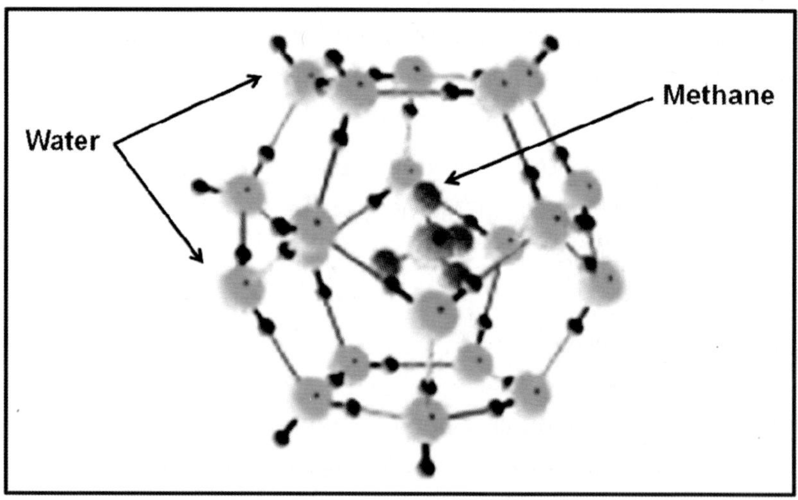

Figure 3-6. Illustration of Methane Hydrate [after USGS Hydrate, 2009]

3.6.2 Tight Gas Sands and Shale Gas

Coal gas, tight gas sands, and shale gas are characterized by low permeabilities. Tight gas sands have permeability less than 0.1 millidarcy. Shale is a low permeability source rock found throughout the world. Permeability in gas shales can be on the order of 0.001 millidarcy (or 1 microdarcy). Gas shale has enough porosity to hold large amounts of gas but low permeability makes it difficult to economically produce hydrocarbons. Gas production from shale has flow mechanisms that are similar to those required for gas production from coal. Shale gas must diffuse through the shale until it reaches a permeable path that allows it to flow to a well. Permeable pathways are typically natural or induced fractures.

The geographic distribution of tight gas sands and gas shale is better understood in the United States than it is elsewhere in the world. Table 3-7 compares the volumes of unconventional gas resources for the United States at the beginning of 2009. These estimates were compiled by the Natural Gas Supply Association (NGSA) from data provided by the United States EIA, the United States Federal Energy Regulatory Commission (FERC), and the United States Potential Gas Committee (PGC). The former two organizations (EIA and FERC) are government agencies and the latter organization (PGC) is a non-profit organization.

Table 3-7
Size of Unconventional Gas Resources in the United States circa January 2009 [NGSA Unconventional Gas, 2009]

Resource	Trillion Cubic Feet	Source
Tight gas	309	US EIA
Shale gas	742	US FERC
Coal gas	163	US PGC

Gas recovery from tight gas sands and shales depends on our ability to produce gas at an economical rate. Advances in drilling and stimulation technologies have made natural gas production from low permeability sands and shale increasingly cost-effective. Two technologies have greatly improved the economics: horizontal drilling and hydraulic fracturing. These technologies are discussed in Chapter 4.

3.7 Global Distribution of Oil and Gas

The global distribution of oil and gas is illustrated by presenting the size of a nation's reserves. Table 3-8 lists fifteen countries with the largest oil reserves and fifteen countries with the largest gas reserves in 2008. National reserves are found by summing the reserves for all of the reservoirs in the nation. Reserves for 2008 are presented in Table 3-8 even though 2009 values were available because the 2008 values can be used in conjunction with 2008 production and consumption values discussed later.

Table 3-8
Nations with Largest Crude Oil and Natural Gas Proved Reserves in 2008
[US EIA website, 2009]

Crude Oil Reserves (billion barrels)		Natural Gas Reserves (trillion cubic feet)	
World	1332.04	World	6212.34
Saudi Arabia	266.75	Russia	1680.00
Canada	178.59	Iran	948.20
Iran	138.40	Qatar	905.30
Iraq	115.00	Saudi Arabia	253.11
Kuwait	104.00	United States	237.73
United Arab Emirates	97.80	United Arab Emirates	214.40
Venezuela	87.04	Nigeria	183.99
Russia	60.00	Venezuela	166.26
Libya	41.46	Algeria	159.00
Nigeria	36.22	Iraq	111.94
Kazakhstan	30.00	Turkmenistan	100.00
United States	21.32	Kazakhstan	100.00
China	16.00	Indonesia	93.90
Qatar	15.21	Malaysia	83.00
Algeria	12.20	China	80.00

The geographic distribution of crude oil and natural gas is essentially the same from one year to the next. The geographic distribution of reserves can change if major new fields are discovered. Offshore West Africa and deepwater Gulf of Mexico are examples of regions where new reserves are being discovered. Other factors such as advances in technology and changes in price can influence the amount of reserves.

Table 3-9 shows a regional distribution of crude oil and natural gas reserves. Eurasian reserves are primarily Russian reserves. The table shows that the Middle East has the largest reserves in both categories and Europe has the smallest reserves.

Table 3-9
Regional Distribution of Crude Oil and Natural Gas Proved Reserves in 2008
[US EIA website, 2009]

Crude Oil Reserves (billion barrels)		Natural Gas Reserves (trillion cubic feet)	
World	1332.04	World	6212.34
Middle East	748.29	Middle East	2548.90
North America	211.56	Eurasia	2014.80
Africa	114.84	Africa	489.63
Central & South America	109.86	Asia & Oceania	415.39
Eurasia	98.89	North America	309.78
Asia & Oceania	34.35	Central & South America	261.80
Europe	14.27	Europe	172.04

3.8 Activities

True-False

Specify if each of the following statements is True (T) or False (F).

1. A hydrocarbon reservoir must be able to trap fluids.
2. The biogenic theory of the origin of fossil fuels depends on the death and decay of life from the surface biosphere.

3. Permeability is needed for fluid to flow in a reservoir.
4. Shale gas is produced from reservoirs with high permeability.
5. Methane hydrate contains more water than methane.
6. "Easy" oil is found at the bottom of the resource pyramid.
7. Hydrocarbon liquids are often classified by gas-to-liquid ratio.
8. The amount of oil and gas reserves in the world does not change.
9. The Middle East has only oil reserves and no natural gas reserves.
10. SPE includes gases in its definition of petroleum.

Questions

1. Petroleum fluids are usually less dense than water and will migrate up through the ground until they encounter impermeable barriers and collect in ______.
2. The ________ theory argues that oil originated as the product of processes deep within the earth's mantle.
3. What are the two primary elemental constituents of petroleum fluids?
4. Which of the following factors are needed to form a hydrocarbon reservoir?

a.	The biogenic theory must be correct
b.	A hydrocarbon source must exist
c.	A flow path must exist between hydrocarbon source and reservoir
d.	A trap must exist for capturing fluids
e.	The timing must be right, that is, the fluid trap must exist at the time that fluid is moving from the hydrocarbon source to the reservoir

5. Which of the following theories is considered the mainstream theory of the origin of fossil fuels: biogenic or abiogenic?
6. How is a gas hydrate formed?
7. Calculate the SPE-PRMS reserves for a mean of 200 MMSTBO and a standard deviation of 40 MMSTBO.
8. How does volatile oil behave when pressure drops?
9. Hydrocarbon liquids can be classified using ____ and viscosity.
10. What is SAGD and where is it used?

CHAPTER 4

PEAK OIL

Oil was the fuel of choice to meet global energy needs at the end of the 20th century. Many experts believe that the supply of oil will reach a peak in the first quarter of the 21st century – sometime between now and 2025 – and will begin to decline. The decline in produced oil will occur despite an increasing demand for energy. The combination of increased demand and reduced supply will lead to a significant increase in the price of oil. According to this scenario, other sources of energy will begin to replace the increasingly expensive oil.

The objective of this chapter is to discuss the evidence that peak oil has been reached. Following Fanchi [2010], we describe the life of an oil and gas reservoir so that we can better understand the factors that affect peak oil. We then present the status of current oil and gas production, and analyze the evidence for peak oil.

4.1 OIL AND GAS PRODUCTION

The production life of the reservoir begins when fluid is withdrawn from the reservoir. Reservoir boundaries are established by seismic surveys and delineation wells. A seismic survey is based on vibrations that are sent into the earth. Part of the vibrational energy is reflected back to the surface when the vibrations encounter rock formations in the subsurface. The reflected energy can be used to infer a picture of the geology of the subsurface. Seismic surveys are typically used to explore for structures that can trap oil and gas. When a prospect is found, wells are drilled to see if hydrocarbon is actually present. A well is a "dry hole" if it does not encounter hydrocarbons. The well that encounters hydrocarbon is the discovery well. Delineation wells are drilled to establish the size of the reservoir, but can also be used for production or injection later in the life of the reservoir.

Production can begin immediately after the discovery well is drilled, or years later after several delineation wells have been drilled. The number of wells used to develop the field, the location of the wells, and their flow characteristics are among the many issues that must be addressed by reservoir management.

4.1.1 History of Drilling Methods

The first method of drilling for oil in the modern era was introduced by Edwin Drake in the 1850's and is known as cable-tool drilling. In this method, a rope connected to a wood beam had a drill bit attached to the end. The beam rocked back and forth so that it lifted and then dropped the bit into the ground. Cable-tool drilling does not work in soft-rock formations where the sides of the hole can collapse. Cable-tool drilling has been largely replaced by rotary drilling.

Developed in France in the 1860's, rotary drilling was first used in the United States in the 1880's because it could drill into the soft-rock formations of the Corsicana oilfield in Texas. Rotary drilling uses a rotating drill bit to penetrate into the earth. Drilling mud is circulated down the drill pipe through nozzles in the drill bit. The mud flows up the annulus between the drill pipe and the wellbore wall. Drilling mud is designed to cool the drill bit and carry rock cuttings up the wellbore to the surface.

Rotary drilling gained great popularity after Captain Anthony F. Lucas drilled the Lucas 1 well at Spindletop, near Beaumont, Texas. The Lucas 1 well was a discovery well. So much oil flowed up the well that it engulfed the drilling derrick and it appeared as if the earth was gushing oil. Instead of flowing at the expected 50 barrels of oil per day, the well produced up to 75,000 barrels per day. The Lucas gusher began the Texas oil boom [Yergin, 1992, pages 83-85]. Since then, rotary drilling has become the primary means of drilling. A modern drilling rig is shown in

Figure 4-1. The rotary mechanism that is used to rotate the drill pipe with the attached drill bit is set in the drill floor.

Figure 4-1. Illustration of a Drilling Rig

Once a hole has been drilled, it is necessary to complete the well. A well is completed when it is prepared for production. The first completed well of the modern era was completed in 1808 when two American brothers, David and Joseph Ruffner, used wooden casings to prevent low-concentration salt water from diluting the high-concentration salt water they were extracting from deeper in their saltwater well [Van Dyke, 1997, pages 145-146]. A modern wellbore is sketched in Figure 4-2.

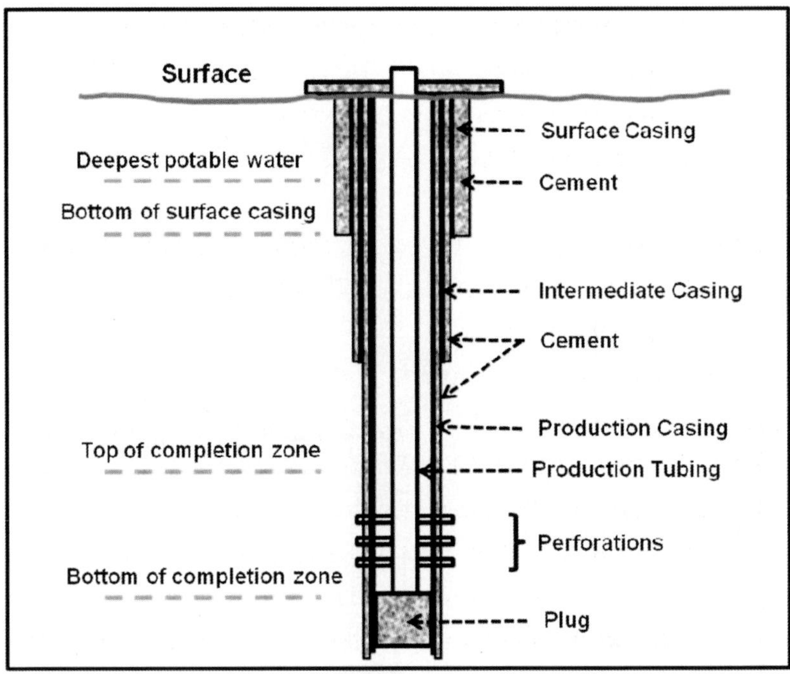

Figure 4-2. Illustration of a Wellbore Diagram for a Vertical Well

Steel casing and cement are typically used to isolate shallow formations from fluid in the wellbore. Cement is injected through the drill pipe and into the annulus between the drill pipe and the bore hole wall. Cement is used to form a barrier between the bore hole and adjacent formations. Leaks in the cement can result in fluid movement from the producing interval to other formations. This can affect production and injection of fluids. Cement bond log measurements can be used to check the integrity of the cement.

Wellbores may be completed using a variety of techniques. A common completion technique is to place tubing adjacent to the productive interval and then use a perforation gun to shoot perforations through the tubing walls, as illustrated in Figure 4-2. Reservoir fluid can flow into the tubing through the perforation holes. A plug is set at the bottom of the completion zone in Figure 4-2 to prevent fluid from flowing into the borehole below the completion zone. In open hole completions, tubing is set above the productive interval and fluid is allowed to fill the open wellbore until it can flow into the tubing and up the wellbore. A wire screen can be placed in the wellbore adjacent to the productive interval in reservoirs where rock fragments or grains can move into the wellbore. This can happen, for example, when an unconsolidated formation is being produced.

It is sometimes necessary to provide energy to extract oil from reservoirs. Oil can be lifted using pumps or injecting gas into the well stream to increase the buoyancy of the gas-oil mixture. The earliest pumps used the same wooden beams that were used for cable-tool drilling. Oil companies developed central pumping power in the 1880's. Central pumping power used a prime mover, a power source, to pump several wells. In the 1920's, the development of a beam pumping system made it feasible to allocate a pump to each well. A beam pumping system is a self-contained unit that is mounted at the surface of each well and operates a pump in the hole. More modern techniques include gas-lift and electric submersible pumps. Energy can also be provided by injecting fluids like water, air, or carbon dioxide into the reservoir. Improved recovery technology includes traditional secondary recovery processes such as waterflooding and immiscible gas injection, as well as enhanced oil recovery (EOR) processes. EOR processes are often classified as follows: chemical, miscible, thermal, and microbial.

4.1.2 Modern Drilling Methods

Advances in drilling technology are extending the options available for prudently managing subsurface reservoirs and producing fossil fuels, especially oil and gas. Modern drilling technologies discussed here include horizontal wells, multilateral wells, and infill drilling.

A well is a string of connected, concentric pipes. The path followed by the string of pipes is called the trajectory of the well. Historically, wells were drilled vertically into the ground and the well trajectory was essentially a straight, vertical line. Today, wells can be drilled so that the well trajectory is curved. A curved wellbore trajectory is possible because the length of each straight pipe that makes up the well is small compared to the total well length. The length of a typical section of pipe in a well is 30 ft to 40 ft, while wells can be thousands of feet long. Wells with one or more horizontal trajectories are sketched in Figure 4-3.

Wells with more than one hole can be drilled. Each hole is called a lateral or branch and the well is called a multilateral well. For example, a bilateral well is a well with two branches. Figure 4-3 shows examples of modern multilateral well trajectories.

A well can begin as a vertical well, and then be modified to a horizontal or multilateral well. The vertical section of the well is called the main (mother) bore or trunk. The point where the main bore and a lateral meet is called a junction. When the vertical segment of the well reaches a specified depth called the kick-off point (KOP), mechanical wedges (whipstocks), or other downhole tools are used to change the direction of the drill bit and alter the well path. The beginning of the horizontal segment is the heel and the end of the horizontal segment is the toe. The distance, or reach, of a well from the drilling rig to final bottomhole location can exceed 6 miles. Wells with unusually long reach are called extended reach wells.

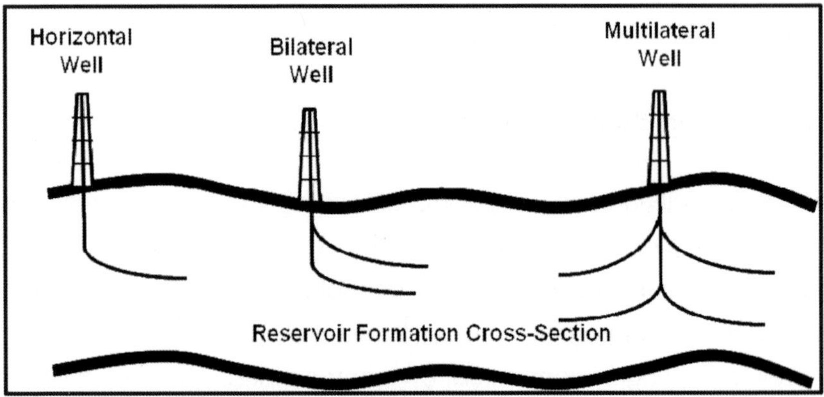

Figure 4-3. Multilateral Wells

Multilateral wells make it possible to connect multiple well paths to a common wellbore. Multilateral wells have many applications. For example, multilateral wells are used in offshore environments where the number of well slots is limited by the amount of space available on a platform. They are also used to produce fluids from reservoirs that have many compartments. A compartment in a reservoir is a volume that is isolated from other parts of the reservoir by barriers to fluid flow such as sealing faults.

Horizontal, extended reach, and multilateral wellbores that follow subsurface formations are providing access to more parts of the reservoir from fewer well locations. This provides a means of minimizing environmental impact associated with drilling and production facilities, either on land or at sea. Extended reach wells make it possible to extract petroleum from beneath environmentally or commercially sensitive areas by drilling from locations outside of the environmentally sensitive areas. In addition, offshore fields can be produced from onshore drilling locations, and reduce the environmental impact of drilling by reducing the number of surface drilling locations.

Horizontal drilling provided significant advantages needed to make unconventional gas production economical. For example, shale layers can be a few hundred feet thick and extend laterally for miles. The Barnett Shale in the United States is up to 450 feet thick and covers an area of over 5,000 square miles. Thus a vertical well, which may pass through 400 feet of shale, can extract gas from a much smaller volume than a horizontal well that could run through two miles or more of shale. Figure 4-4 illustrates this difference. The figure shows that a horizontal well can intersect more of the reservoir when the areal extent of the reservoir is substantially larger than the thickness of the reservoir. The orientation of fractures depends in part on the orientation of the well, and also on the rock properties of the reservoir.

Multiple wells drilled from a single well pad decrease the need for large numbers of potentially noisy and aesthetically unappealing surface well sites. With multi-well drilling sites, the number of locations that must have transport pipelines and that must be operated and maintained is minimized, saving time and money for operators. Horizontal drilling makes it possible to locate a well site outside of a highly populated area and drill underneath the population without being seen or heard. This development has further increased interest in shale gas because a large volume of shale gas is found under large cities. Urban drilling had largely become unacceptable given the unsightliness and potential health hazards of drilling rigs. Horizontal drilling increases public support for urban drilling.

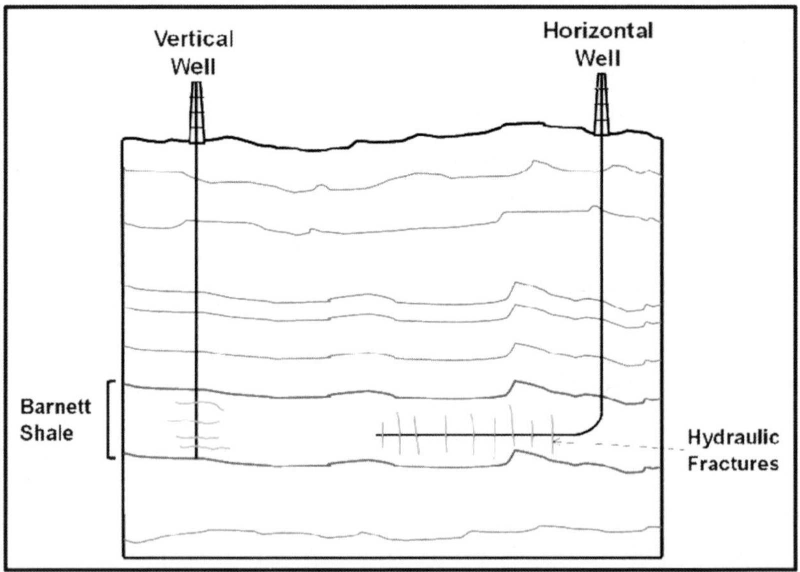

Figure 4-4. Schematic of Barnett Shale Cross Section

4.1.3 Hydraulic Fracturing

Wells drilled in low permeability rock often need to be stimulated. One key stimulation technology is hydraulic fracturing. In the 1980s and 1990s, developments in stimulation technology made it possible to create cracks in rock using a process called hydraulic fracturing (or "fracing," pronounced "frack-ing"). The fracturing process involves the injection of liquid slurry at a pressure that is large enough to break the rock. Once fractures have been created in the formation, a proppant such as manmade pellets or coarse grain sand is injected into the fracture to prevent the fracture from closing, or healing, when the injection pressure is discontinued. The proppant keeps the fracture open enough to provide a higher permeability flow path for gas to flow to the production well. Today, fracing is used to improve oil and gas flow rates from low permeability rock. Modern applications include oil shale, gas shale, and tight gas sands.

The combination of hydraulic fracturing and horizontal drilling technologies was the catalyst for the growth of the shale gas production industry. In 2002, Mitchell Energy was among the first companies to implement horizontal drilling in a shale formation, doing so in the Barnett Shale in North Texas. Hydraulic fracturing had been in place for years, but the results had limited success. Horizontal drilling made it possible for horizontal wellbores to intersect fractures in a much larger volume of the reservoir than vertical drilling. The combination of hydraulic fracturing and horizontal drilling increased the rate that gas could be produced from targets like gas-rich, low permeability shale.

4.1.4 Production System

A production system can be thought of as a collection of subsystems illustrated in Figure 4-5. Fluids are taken from the reservoir using wells. Wells must be drilled and completed. An example of a drilling rig on an offshore platform in dry dock is shown in Figure 4-6. The performance of the well depends on the properties of the reservoir rock, the interaction between the rock and fluids in the reservoir, and properties of the fluids in the reservoir. Reservoir fluids include the fluids originally contained in the reservoir as well as fluids that may be introduced as part of the reservoir management process. Well performance also depends on the properties of the well itself, such as its cross-section, length, trajectory, and completion. The completion of the well establishes the connection between the well and the reservoir, as described in Section 4.1.1 above.

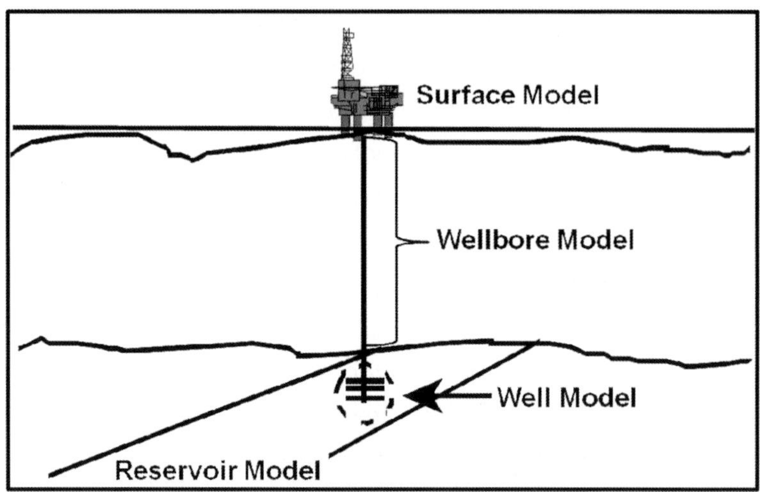

Figure 4-5. Production System

Figure 4-6. Offshore Platform in Dry Dock, Galveston, Texas (Fanchi, 2003)

Surface facilities are needed to drill, complete and operate wells. Drilling rigs may be moved from one location to another on trucks, ships, or offshore platforms; or drilling rigs may be permanently installed at specified locations. The facilities may be located in desert climates in the

Middle East, stormy offshore environments in the North Sea, arctic climates in Alaska and Siberia, and deepwater environments in the Gulf of Mexico and off the coast of West Africa.

4.1.5 Processing and Transport

Produced fluids must be recovered, processed and transported to storage facilities and eventually to the consumer. Processing can begin at the well site where the produced wellstream is separated into oil, water and gas phases. Further processing at refineries separates the hydrocarbon fluid into marketable products (Figure 4-7). A typical barrel of crude oil is used to produce gasoline, diesel, jet fuel, liquefied petroleum gas, heavy fuel oil, and other products.

Figure 4-7. A South Texas Refinery (Fanchi, 2002)

Many modes of transportation are used to transport oil and gas. They include pipelines, tanker trucks, double hulled tankers, and ships capable of carrying liquefied natural gas. The mode of transport depends on such factors as safety, distance, and state of the fluid. The transport distance can be a few miles from field to processing facility, or thousands of miles from refinery to consumer.

4.1.6 Power Plants

Fossil fuels were used throughout the 20th century to drive turbines in electricity generating power plants. Combustion of fossil fuels, especially coal, emits undesirable gases. One way to reduce this problem is to use cleaner natural gas in gas-fired power plants.

A conventional gas-fired power plant is the combined cycle power plant. It uses heat from combustion in more than one electricity generating cycle. The primary cycle uses hot gases from combustion to drive a gas turbine. Waste heat, typically in the form of exhaust, is captured and used to create steam, which is then used to drive a steam turbine. Both the gas turbine and the steam turbine generate electricity.

An alternative configuration to the combined cycle configuration is obtained by using waste heat from the gas turbine for direct heating. The direct heating configuration is known as cogeneration and uses waste heat for space heating or water heating. Both the combined cycle configuration and the cogeneration configuration improve the efficiency of energy use.

Gas-fired power plants are cleaner than coal-fired power plants, but they still emit some greenhouse gas, notably carbon dioxide. Another variation of the gas-fired combined cycle power plant is to attach a carbon capture and sequestration (CCS) system. An example of a CCS system is a system that captures carbon dioxide from the gas combustion process and injects the captured carbon dioxide into a geological reservoir. Gas storage in a geological reservoir is known as geological sequestration.

4.2 Global Oil and Gas Production and Consumption

The global distribution of oil and gas production and consumption is illustrated by presenting the leading nations in production and consumption categories. Table 4-1 lists the five countries with the largest production and consumption of oil in 2008. Table 4-2 lists the five countries with the largest oil exports and imports in 2008.

Table 4-1
Top 5 Oil Producing and Consuming Nations in 2008
[US EIA website, accessed November 2009]

Producing Nations (million barrels per day)		Consuming Nations (million barrels per day)	
Saudi Arabia	10.8	United States	19.5
Russia	9.8	China	7.8
United States	8.5	Japan	4.8
Iran	4.2	India	3.0
China	3.0	Russia	2.9

Table 4-2
Top 5 Oil Exporting and Importing Nations in 2008
[US EIA, accessed November 2009]

Exporting Nations (million barrels per day)		Importing Nations (million barrels per day)	
Saudi Arabia	8.4	United States	11.0
Russia	6.9	Japan	4.6
United Arab Emirates	2.5	China	3.9
Iran	2.4	Germany	2.4
Kuwait	2.4	Korea, South	2.1

Table 4-3 lists the five countries with the largest production and consumption of dry natural gas in 2008. Table 4-4 lists the five countries with the largest dry natural gas exports and imports in 2008. The United

States was one of the top 5 producers of oil and dry natural gas in 2008, and the leading consumer and importer of oil and dry natural gas.

Table 4-3
Top 5 Dry Natural Gas Producing and Consuming Nations in 2008 [US EIA, accessed November 2009]

Producing Nations (billion cubic feet per day)		Consuming Nations (billion cubic feet per day)	
Russia	64.1	United States	63.6
United States	56.3	Russia	46.0
Canada	16.5	Iran	11.5
Iran	11.3	Japan	9.8
Norway	9.6	United Kingdom	9.3

Table 4-4
Top 5 Dry Natural Gas Exporting and Importing Nations in 2008 [US EIA, accessed November 2009]

Exporting Nations (billion cubic feet per day)		Importing Nations (billion cubic feet per day)	
Canada	10.0	United States	10.9
Norway	9.2	Japan	9.2
Netherlands	6.0	Germany	8.9
Algeria	5.8	Italy	7.4
Qatar	5.5	Ukraine	6.2

With the discovery of drilling methods capable of accessing unconventional natural gas domestically, the United States has shifted its focus for power plants from coal towards clean, dry natural gas. While this process has been accelerating over the past decade or so, the United States still needs more natural gas than it can produce since the vast shale gases in the country are just barely being tapped. It is likely that the need to import dry natural gas in the United States will decrease as more domestic gas production comes on line.

4.3 The First Oil Crisis

One of the factors used to decide which energy source to use is price. To understand the current oil price, we must examine past events that have shaped the global oil market. While we could easily dedicate an entire chapter to the events that have impacted the way oil functions today, we chose to focus on an event that has served as the catalyst for many future events. This event is known as the first oil crisis, and was set in motion by the ongoing conflict in the Middle East.

The Middle East has been embroiled in conflict since the formation of Israel in 1947. The British occupied Palestine since 1917, which was near the end of World War I. After World War I ended in 1918, the League of Nations approved the Mandate of Palestine, which authorized British administration of Palestine.

The United Nations partitioned Palestine into a Jewish state, an Arab state, and a UN administered Jerusalem when World War II ended in 1945. The partitioning took effect in November, 1947 and the British left Palestine in 1948. The partitioning motivated two conflicting responses. Arab leaders rejected the partitioning, and Israel declared independence and asserted its sovereignty when Britain left the region in 1948. The initial result of the partitioning was a civil war between Palestinian Jews and Palestinian Arabs. When Israel declared itself a sovereign state, the civil war escalated to an international war between Israel and nearby Arab countries. The war ended in July 1949, but it was not resolved. The conflict in the Middle East set the stage for a future war that would launch the first oil crisis.

4.3.1 Events Leading Up to the Crisis

On September 14, 1960, the Organization of the Petroleum Exporting Countries (OPEC) was founded at a conference in Baghdad, Iraq. Thirteen countries including Iran, seven Arab countries, Indonesia, Angola, Nigeria, Ecuador and Venezuela joined the group with the goal of creating a neutral bargaining unit for Third World oil producing countries. At the time, these countries were facing pressure from oil companies to lower prices and had little leverage individually.

Initially OPEC simply sought to increase the share of oil revenues given to the oil producing countries and to increase their control over the supply of oil. By the start of the next decade, OPEC had become a unified force and was gaining influence over world oil supply and price.

In May, 1967 Egypt expelled United Nations forces from the Sinai region separating Egyptian and Israeli territory. Israel, threatened by the presence of Egyptian forces on their border, called up reservists in preparation for armed conflict. On June 5, 1967, Israel attacked the Egyptian forces. Jordan and Syria joined Egypt in the struggle against Israel. The war lasted from June 5 to June 10, 1967 and is called the Six-Day War. Israel won a decisive victory and tripled its size by taking control of the Sinai Peninsula, the Gaza Strip, the West Bank of the Jordan River including East Jerusalem, and the Golan Heights.

Four years later, in a seemingly unrelated event, the United States and several other countries decided in 1971 to discontinue using gold as a standard for the value of their currencies. This led directly to depreciation of the value of the dollar and other currencies that oil producing nations were receiving for their goods.

The new currency valuation system had an adverse effect on the purchasing power of Third World oil producing countries. Change in the value of currencies was accompanied by price increases in goods sold to these countries from the major oil consumers (goods such as wheat, grain, sugar, cement, even refined petroleum products) while the price of oil remained the same. Oil producing countries, including the members of OPEC, were receiving less real income for their exported oil at the same time that they were paying higher prices for the goods they purchased. Members of OPEC decided to begin pricing oil against the value of gold instead of the value of the dollar. The result was the first stage of the "oil shock" felt in the U.S. and throughout the world as oil prices suddenly increased.

Meanwhile, in the Middle East, tensions remained high as Arab nations viewed the existence of the Israeli state as an affront to the region. The Egyptians and Syrians, beaten in the Six-Day War, sought to recover the land they had lost as well as to defend their national honor after the humiliation suffered in the defeat in 1967. On October 6, 1973, on the Jewish holiday of Yom Kippur, Egypt and Syria attacked Israel in what became known as the Yom Kippur War.

Israel's closest ally, the United States, expected the war to be short-lived based on the length of previous wars, but the Egyptians and Syrians were receiving assistance from Arab allies and support from the U.S.S.R. In the first 24-48 hours of the war, the attacking nations made positive advances, but shortly thereafter momentum began to shift to the defending Israelis, although losses were unexpectedly high.

Concerned that Israel was suffering significant losses, United States Secretary of State Henry Kissinger and President Richard Nixon initiated Operation Nickel Grass, an airlift operation designed to bring supplies to Israel to help in the war. There were also reports that the United States had flown reconnaissance missions over Egypt to provide intelligence to Israel about Egyptian military movements and placements in preparation for an Israeli counterattack, launched on October 14, 1973.

The United States' involvement in the Yom Kippur War would turn out to have enormous implications. Virtually the entire Arab world considered Israel an adversary and sought its destruction. On September 1, 1967, at the Khartoum Arab Summit, participating Arab nations signed the Khartoum Resolution which included the infamous "three no's": no peace with Israel, no recognition of Israel, and no negotiation with Israel. By supporting Israel in the war, the U.S. not only enraged Egypt and Syria, but the entire Arab world. The final straw came on October 17 when President Nixon requested $2.2 billion in emergency aid from Congress for Israel. Two days later, on October 19, 1973, thirteen days after the start of the Yom Kippur war, Libya initiated an oil embargo against the United States and the rest of the Arab nations followed suit.

4.3.2 The Arab Oil Embargo

The purpose of the oil embargo was two-fold. First, it was important that Arab countries supporting the war against Israel provide an incentive for the United States to withdraw its support from the Israeli side. By eliminating the supply of oil to the U.S. from OPEC members, the price of oil for the U.S. and the price of oil products within the U.S. were expected to increase. The second purpose of the embargo was to punish the United States for taking sides in the conflict, and particularly for taking the side of Israel. The U.S. believed it had little choice in the matter, however, since the U.S. and Britain were largely responsible for the existence of the current Israeli state. In addition, the U.S. knew that the U.S.S.R. was supporting the Arab side of the conflict, effectively making the Yom Kippur War a battlefield in the Cold War.

The Yom Kippur War ended a week after the oil embargo began on October 19, 1973, but the embargo did not end until March 17, 1974. The ceasefire for the war was largely negotiated by the U.S. and U.S.S.R. on October 22, although the fighting continued for several days and skirmishes continued even longer.

The impact of the oil embargo was exacerbated by the collapse of the United States stock market in January 1973. The 1973-74 market crash was in large part a result of the destabilization of the United States' currency following Nixon's decision to move the country off the gold standard. The price of a barrel of crude oil from Saudi Arabia, Iran or Libya increased by as much as 400% from January 1973 to January 1974. Figure 4-8 shows crude oil prices from these countries from 1970 to 2009, using nominal (not adjusted for inflation) dollars. Americans were forced to begin rationing gasoline since gasoline was suddenly scarcer domestically than it had been since World War II. Arab leaders learned that oil could be used as leverage over the economy of a global superpower, and U.S. leaders realized that the energy destiny of the United States had become dependent on the actions of other nations. The first oil crisis is an important modern example of the interconnection between politics, economics and energy.

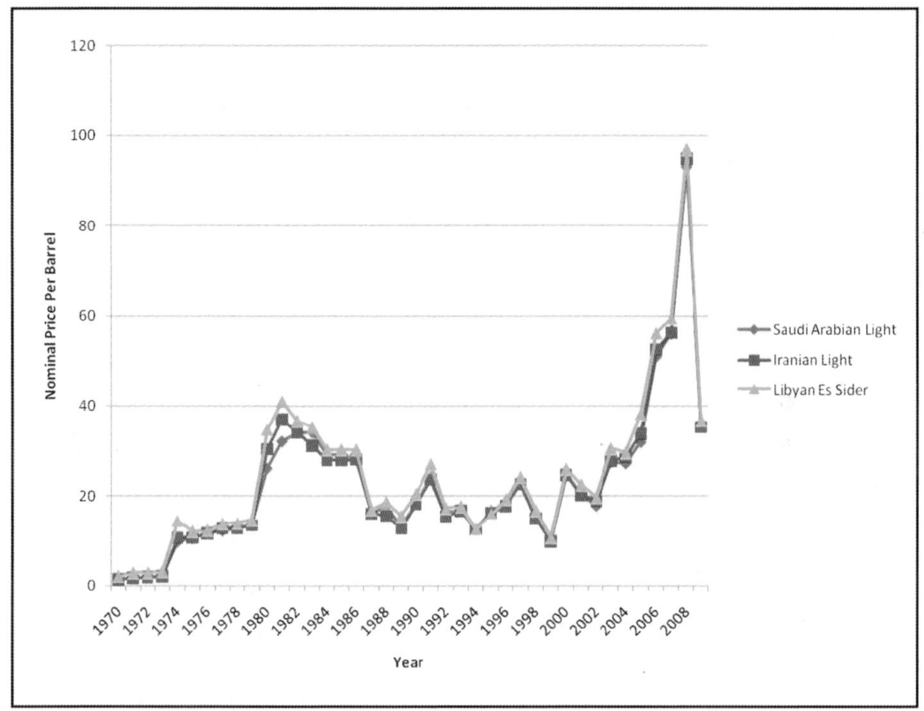

Figure 4-8. Crude Oil Price from 1970-2009

4.4 The Price of Oil

Since the first oil crisis in 1973, alarmists have made dire predictions in the media that the price of oil will increase with virtually no limit. These predictions neglect market forces that constrain the price of oil and other fossil fuels. Historical events provide insight into the sensitivity of oil price.

The effects of the Arab oil embargo were felt immediately. From the beginning of 1973 to the beginning of 1974, the price of a barrel of oil more than doubled. Americans were forced to ration gasoline, with lines at gas stations accompanied by price gouging. The Arab oil embargo prompted nations around the world to begin seriously considering a shift away from a carbon-based economy.

4.4.1 How Does Oil Price Affect Oil Recovery?

Many experts believe we are running out of oil. It is becoming increasingly difficult to discover new reservoirs that contain large volumes of oil and gas. Much of the exploration effort is focusing on less hospitable climates, such as arctic conditions in Siberia and deep water, offshore regions near West Africa. Yet we already know where large volumes of oil remain: in the reservoirs that have already been discovered and developed. Current development techniques have recovered approximately one third of the oil in known fields. That means roughly two thirds remains in the ground where it was originally found.

The efficiency of oil recovery depends on cost. Companies can produce much more oil from existing reservoirs if they are willing to pay for it, and if the market will support that cost. Most oil producing companies choose to seek and produce less expensive oil so they can compete in the international marketplace. Table 4-5 illustrates the sensitivity of oil producing techniques to the price of oil. Oil prices in the table represent an inflation rate of 10% relative to 1997 prices presented in the first edition of this book. The inflation rate for oil prices depends on a number of factors, such as size and availability of supply and demand.

Table 4-5
Sensitivity of Oil Recovery Technology to Oil Price

Oil Recovery Technology	Oil Price (US$ per barrel in year 2009 US$)
Conventional	40 – 80
Enhanced Oil Recovery (EOR)	60 – 130
Extra Heavy Oil (e.g. tar sands)	60 – 140
Alternative Energy Sources	60 – 190

Table 4-5 shows that more sophisticated technologies can be justified as the price of oil increases. It also includes a price estimate for alternative energy sources, such as wind and solar. The lower price in the price range is due to wind energy, which has become economically competitive with oil as an energy source for generating electricity. In some cases there is overlap between one technology and another. For example, steam flooding is an Enhanced Oil Recovery (EOR) process that can compete with conventional oil recovery techniques such as waterflooding, while chemical flooding is an EOR process that can be as expensive as many alternative energy sources.

4.4.2 How High Can Oil Prices Go?

In addition to relating recovery technology to oil price, Table 4-5 contains another important point: the price of oil will not rise without limit. For the data given in the table, we see that alternative energy sources become cost competitive when the price of oil rises above US$60 per barrel. If the price of oil stays at US$60 per barrel or higher for an extended period of time, energy consumers will begin to switch to less expensive energy sources. This switch is known as product substitution and has already begun in many countries. For example, consumers in European countries pay much more for gasoline than consumers in the United States. Countries such as Denmark, Germany and Holland are rapidly developing wind energy as an alternative to fossil fuels. France has opted to rely on nuclear fission energy.

Historically, we have seen oil exporting countries try to maximize their income and minimize competition from alternative energy and expensive oil recovery technologies by supplying just enough oil to keep the price at around US$60 per barrel. Oil importing countries can attempt to minimize their dependence on imported oil by developing technologies that reduce the cost of alternative energy. If an oil importing country contains mature oil reservoirs, the development of relatively inexpensive technologies for producing oil remaining in mature reservoirs or the imposition of economic incentives to encourage domestic oil production can be used to reduce the country's dependence on imported oil.

Point to Ponder: Why would a country want to keep reserves of fossil fuels secret?

Every country depends to some extent on fossil fuels. As we have seen, fossil fuels are an important fuel source for commercial, industrial, and transportation systems. The amount of fossil fuels in a country has an impact on that country's ability to function as a geopolitical power. A country may choose to keep the size of its resource base secret to limit information to its adversaries and introduce uncertainty in negotiations with businesses and other countries.

Recent history has illustrated why some countries would want to mislead people about their fossil reserves. Knowing that fossil fuels are presumed to be non-renewable, this resource has high value to many countries that have few other resources of value. Keeping their actual reserves secret lets countries, such as those in the Middle East, control the supply and price of their resources as long as possible, and gives them a bargaining advantage over foreign nations seeking those resources.

In the late 1980s, members of OPEC had an incentive to mislead others about their national reserves. At the time, OPEC decided to tie production from member countries to their stated reserves. Thus, if a country claimed it had greater reserves, it would be allowed to produce more oil and generate greater revenue. Kuwait was among the first, and the other OPEC nations followed suit, to claim their reserves were suddenly much greater than previously reported. Looking at the numbers, it could be argued that these countries simply switched from reporting their reserves to reporting their total resources which is typically a significantly larger volume. This move inflated their reported reserves, and could impact predictions relating to a future oil peak. [Campbell, 2005]

Many of the "facts" relating to oil and gas reserves internationally are simply estimates based on the amount that each country chooses to report. There is no legal way for a third party to verify the reports made by many countries. Reported reserves can be compared to previous reports and estimates to determine if they are reasonable, but this has limited value. It is this uncertainty that makes it difficult for anyone to accurately predict how much oil and gas remains in the world, and when it might run out. Much of the information must be taken at face value from countries that have a motivation to mislead outsiders.

Further, countries are not the only entities that have an interest in keeping their reserves secret, or at least misleading outsiders about their reserves. In 2004, Royal Dutch/Shell took a hit in its stock market value when it was discovered that the company had been overstating its proved reserves by including fields that had not yet been fully analyzed. These reserves were shifted to the "probable" category or even a lower category, meaning that their value to the firm decreased measurably. Why did Shell overstate the proved reserves? Simply, because it made the company's balance sheet look better. An improvement in the balance sheet can make the company look more valuable, can lead to an increase in the company's stock price, and can improve the company's apparent financial position for receiving loans.

The implications of misleading reserves resonated throughout the industry. It cast doubts on the integrity of reserves reports by other companies, including national oil companies. Shell's top executive resigned in the wake of the scandal, and brought the company under increased scrutiny by the United States Securities and Exchange Commission (SEC).

4.5 Hubbert's Oil Supply Forecast

The emergence of an energy mix is motivated by environmental concerns and by the concern that oil production is finite and may soon be coming to an end. M. King Hubbert studied the production of oil, a non-renewable resource, in the contiguous United States (excluding Alaska and Hawaii) as a non-renewable resource. Hubbert [1956] found that oil production in this limited geographic region could be modeled as a function of time. The annual production of oil increased steadily until a maximum was reached, and then began to decline as it became more difficult to find and produce. The maximum oil production is considered a peak. Hubbert used his method to predict peak oil production in the contiguous United States, which excludes Alaska and Hawaii. Hubbert predicted that the peak would occur between 1965 and 1970. Hubbert then used the methodology that he developed for the contiguous United States to predict the peak of global oil production. He predicted that global oil production would peak around 2000 at a peak rate of 12 to 13 billion barrels/year or approximately 33 to 36 million barrels/day (Figure 4-9).

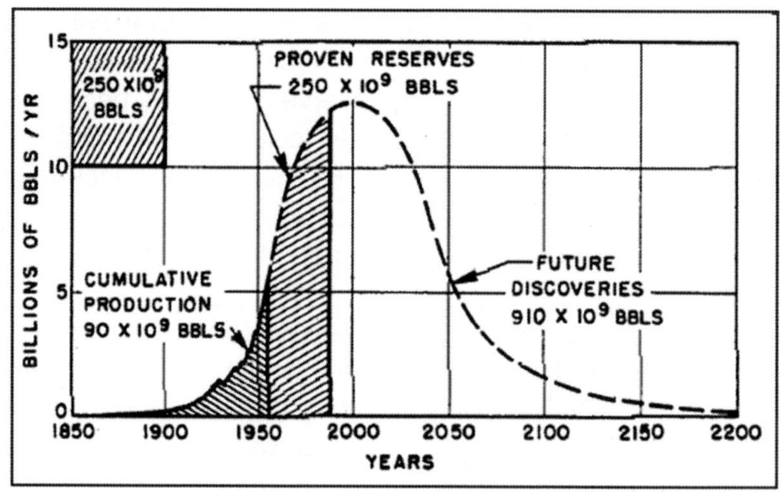

Figure 4-9. Hubbert's Global Oil Prediction [Hubbert, 1956, Figure 20]

Crude oil production in the contiguous United States peaked at 9.4 million barrels per day in 1970. A second peak for the United States occurred in 1988 when Alaskan oil production peaked at 2.0 million barrels per day. The second peak is not considered the correct peak to compare to Hubbert's prediction because Hubbert restricted his analysis to the production from the contiguous United States. Many modern experts consider the 1970 oil peak to be a validation of Hubbert's methodology, and have tried to apply the methodology to global oil production. Analyses of historical data using Hubbert's methodology typically predict that world oil production will peak in the first quarter of the 21st century.

4.5.1 World Oil Production Rate Peak

Forecasts based on an analytical fit to historical data can be readily prepared using publicly available data. Figure 4-10 shows a fit of world oil production rate (in millions of barrels per day) from the United States EIA database. Two fits are displayed. The fit labeled "Gaussian Fit 2000" uses data through year 2000, while the fit labeled "Gaussian Fit 2008" uses data through year 2008. Each fit was designed to match the most recent part of the production curve most accurately. Both fits give oil production rate peaks between 2010 and 2015.

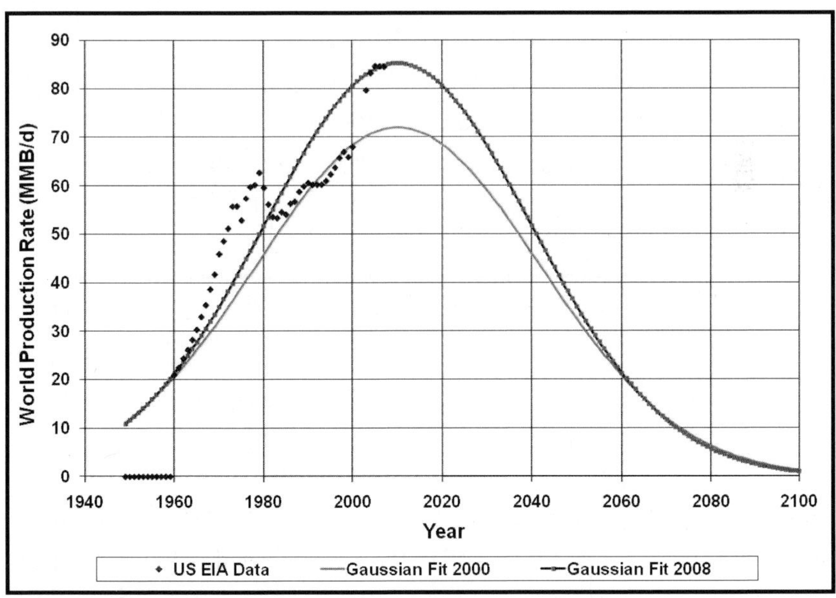

Figure 4-10. World Oil Production Rate Forecast Using Gaussian Curve

The significant increase in actual world oil production rate between 2000 and 2010 is due to a change in infrastructure capacity in Saudi Arabia. This period coincided with a significant increase in oil price per barrel, which justified an increase in facilities needed to produce, collect, and transport an additional one to two million barrels per day of oil.

4.5.2 World Per Capita Oil Production Rate Peak

We saw in Section 4.5.1 that the evidence for a peak in world oil production rate is inconclusive. On the other hand, suppose we consider world per capita oil production rate, which is annual world oil production rate divided by world population for that year. Figure 4-11 shows world per capita oil production rate (in barrels of oil produced per day per person) for the period from 1960 through 2000.

Figure 4-11 shows two peaks in the 1970's. The first peak was at the time of the first oil crisis, and the second peak occurred when Prudhoe Bay, Alaska oil production came online. World per capita oil production rate has been relatively constant since the end of the 1970's, and is significantly below the peak. It appears that world per capita oil production rate peaked in the 1970's. It also appears from Figure 4-11 that oil production rate is managing to keep pace with the increase in world population. We do not know if another, higher peak is possible.

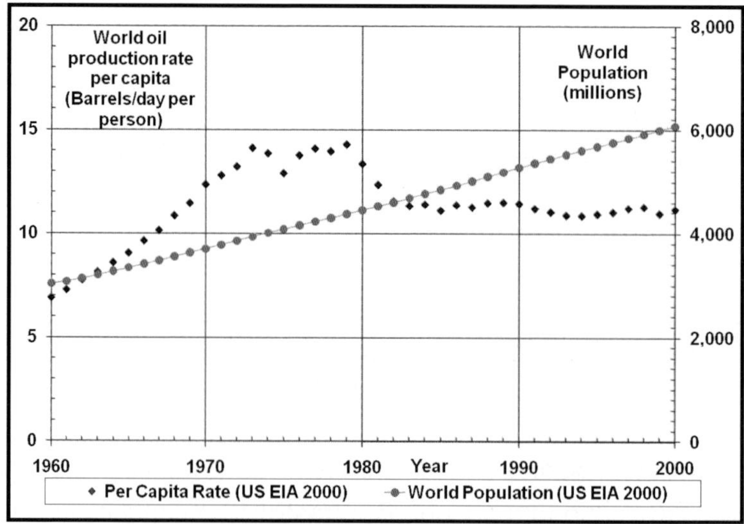

Figure 4-11. World Per Capita Oil Production Rate

4.6 ENVIRONMENTAL ISSUES

Modern strategies of managing oil and gas fields are typically expected to protect the environment. This is due to factors such as government regulation, a desire by producers to be good stewards of the earth, and a strategy to recognize the economic interests of a broad group of stakeholders. Environmental issues associated with the production of conventional fossil fuels include handling naturally occurring radioactive materials (NORM), minimizing emission of pollutants into the environment, and preventing spills. The following examples illustrate some of the issues.

The Louisiana Offshore Oil Production (LOOP) facility is designed to keep the transfer of hydrocarbons between pipelines and tankers away from sensitive coastal areas. Periodic water sampling of surface and produced waters can assure that fresh water sources are not contaminated. In addition, periodic testing for the excavation or production of NORM helps assure environmental compliance and makes it possible for operators to collect and properly dispose of NORM.

Environmental issues associated with gas production from shale gas include meeting the extensive demand for water to conduct hydraulic fracture treatments, disposing produced water containing pollutants, controlling the emanation of toxic chemicals from storage tanks, and preventing the occurrence of seismic events when disposing produced water. Urban operations require controlling truck traffic, acquiring access to rights-of-way, mitigating noise and visual pollution from operations, obtaining well pad locations and reclaiming the environment when the well site is abandoned.

Operating a field with prudent consideration of environmental issues can pay economic dividends. In addition to improved public relations, sensitivity to environmental issues can minimize adverse environmental effects that may require costly remediation and financial penalties. Remediation often takes the form of expensive cleanup operations such as those required after the oil spill from the grounding of the 1989 Exxon-

Valdez oil tanker in Alaska or the 2010 explosion and sinking of the BP Deepwater Horizon offshore platform in the Gulf of Mexico.

Technologies are being developed to improve our ability to clean up environmental pollutants. For example, bioremediation uses living microorganisms or their enzymes to accelerate the rate of degradation of environmental pollutants. Other methods of handling spills include skimming to collect the surface oil, burning the oil, or allowing some types of spilled oil to evaporate naturally.

Failure to develop an environmental management plan that accounts for the interests of all stakeholders, including the public, can lead to unexpected consequences. For example, in 1995 Shell UK reached an agreement with the British government to dispose an oil storage platform called the Brent Spar in the deep waters of the Atlantic. The environmental group Greenpeace led a public challenge to the plan by occupying the platform and supporting demonstrations that, in some cases, became violent. Greenpeace and its allies were concerned that oil left in the platform would leak into the Atlantic. Shell UK abandoned the plan and eventually found another use for the structure as a ferry quay. Governments throughout Europe changed their rules regulating disposal of offshore facilities.

As noted earlier, fossil fuels are commonly consumed by burning. A key environmental issue associated with combustion is the emission of pollutants into the atmosphere. Efforts are underway to minimize the release of gases that are the by-product of hydrocarbon combustion because of their impact on climate change. This is discussed in more detail later.

4.7 ACTIVITIES

True-False

Specify if each of the following statements is True (T) or False (F).

1. Oil wells can be drilled vertically and horizontally.
2. Gas wells have to be drilled vertically.

3. Seismic measurements use vibrations to provide images of the subsurface.
4. Carbon dioxide can be used to increase oil recovery in some oil reservoirs.
5. Rotary drilling was developed in France.
6. Oil recovery does not depend on price.
7. Wells with unusually long extent are called extended reach wells.
8. The United States produces more dry natural gas than it consumes.
9. The Yom Kippur War was the catalyst for the Arab oil embargo on the United States.
10. M. King Hubbert predicted that world oil production would peak between 1965 and 1970.

Questions

1. What role does injected sand play in hydraulic fracturing?
2. What role does cement play in the completion of a wellbore?
3. When a vertical well is modified to become a horizontal or multilateral well, what is the vertical portion called?
4. How does a natural gas-fired power plant use cogeneration?
5. What is CCS?
6. How did the United States' decision to remove its currency from the gold standard affect the oil market?
7. What were two reasons for imposing an 1973 Arab oil embargo on the U.S.in 1973?
8. Why do oil companies want to prevent oil prices from getting "too high"?
9. Why would a country want to keep its oil and gas reserves secret?
10. Why has oil production increased so significantly since 2005?

CHAPTER 5

NUCLEAR ENERGY

Fossil fuels were the dominant energy source in the 20th century. Concern that the supply of fossil fuels is limited and fossil fuel combustion produces greenhouse gases is motivating the search for other sources of energy. One energy source that is receiving renewed interest is nuclear fission energy. Nuclear fission is the process in which a large, unstable nucleus splits into two smaller fragments. Energy from nuclear fission generates heat, which is typically used to generate electric power. Energy consumed by the United States in 2008 included over 8% nuclear fission energy in 2008 (Appendix B). World energy consumption included almost 6% nuclear fission energy in 2006 (Appendix C).

Nuclear energy can be provided when large nuclei split into smaller fragments in the nuclear fission process. Energy can also be provided by combining, or fusing, two small nuclei into a single larger nucleus in the nuclear fusion process. Nuclear fusion reactions are the source of energy supplied by the sun.

Both nuclear fission and nuclear fusion reactions release large amounts of energy. Significant quantities of energy need to be controlled. Some of the energy is waste heat and needs to be dissipated. Some of the energy is useful energy and needs to be transformed into a more useful form. Decay products from the fission process can be highly radioactive for long periods of time and need to be disposed in an environmentally acceptable manner. On the other hand, the byproducts of the fusion process are relatively safe. The purpose of this chapter is to discuss the history of nuclear energy, the role of nuclear energy in the current energy mix, and the potential value of nuclear energy in a future energy mix. We begin by presenting the history of nuclear energy.

5.1 History of Nuclear Energy

Nuclear energy became an important contributor to the global energy mix in the latter half of the 20th century. We obtain nuclear energy from two principle types of reactions: fission, and fusion. Fission is the splitting of one large nucleus into two smaller nuclei; fusion is the joining of two small nuclei into one large nucleus. In both reactions, significant amounts of energy can be released. Nuclear energy in the past and present energy mix has been the result of nuclear fission reactions. Nuclear fusion is a future technology. We consider here the historical development of nuclear energy.

5.1.1 Discovery of the Nucleus

German physicist Wilhelm Roentgen observed a new kind of radiation called x-rays in 1895. Roentgen's x-rays could pass through a human body and create a photograph of the person's internal anatomy. Frenchman Henri Becquerel discovered radioactivity in 1896 while looking for Roentgen's x-rays in the fluorescence of a uranium salt. French physicist and physician Marie Curie (from Warsaw, Poland) and her husband Pierre Curie were the first to report the discovery of a new radioactive element in 1898. They named the element polonium after Marie's homeland.

Ernest Rutherford identified the "rays" emitted by radioactive elements and called them alpha, beta, and gamma rays. Today we know that the alpha ray is the helium nucleus, the beta ray is an electron, and the gamma ray is an energetic photon. By 1913, Rutherford and his colleagues at the Cavendish Laboratory in Cambridge discovered the atomic nucleus by using alpha particles to bombard thin metallic foils. Scattering of alpha particles by atoms in the foil showed that most of the atomic mass is located in the center of the atom. The large mass inside the atom is the nucleus. The constituents of the nucleus were identified as the proton and a new, electrically neutral particle, the neutron. James Chadwick discovered the neutron in 1932 while working in Rutherford's laboratory.

The proton and neutron are classified as nucleons, or nuclear constituents. The number of protons in the nucleus is the atomic number of the nucleus. The mass number of the nucleus is the sum of the number of protons and the number of neutrons.

In an electrically neutral atom, the number of negatively charged electrons is equal to the number of positively charged protons. Isotopes are nuclei with the same atomic number, but different numbers of neutrons. Carbon-12 is the isotope of carbon with a mass number of 12 (six protons plus six neutrons). The trio of isotopes that is important in nuclear fusion is the set of isotopes of hydrogen: hydrogen with a proton nucleus, deuterium with a deuteron nucleus (one proton and one neutron) and tritium with a triton nucleus (one proton and two neutrons).

The number of protons in the nucleus of an element is used to classify the element in terms of its positive electric charge. Isotopes are obtained when electrically neutral neutrons are added to or subtracted from the nucleus of an element. Changes in the number of neutrons occur by several mechanisms. Nucleus changing mechanisms include nuclear emission of helium nuclei (alpha particles), electrons (beta particles), or highly energetic photons (gamma rays). An isotope is said to decay radioactively when the number of protons in its nucleus changes. Elements produced by radioactive decay are called decay products.

The energy that is used to hold the nucleus together is called binding energy. Some binding energy is released when the nucleus splits into two fragments. This splitting, or fission, of the nucleus can occur when the mass number is sufficiently large. The separation of a large nucleus into two comparable fragments is an example of spontaneous fission. The original large nucleus is called the parent nucleus, and the fission fragments are called daughter nuclei. By contrast, the fusion process releases energy when two light nuclei with very small mass are combined to form a larger nucleus.

5.1.2 Radioactivity

Several measures of radioactivity exist. The radiation dose unit that measures the amount of radiation energy absorbed per mass of absorbing material is called the Gray (Gy). One gray equals one Joule of radiation energy absorbed by one kilogram of absorbing material.

A measure of radiation is needed to monitor the biological effects of radiation for different types of radiation. The measure of radiation is the dose equivalent. Dose equivalent is the product of the radiation dose times a qualifying factor. The qualifying factor indicates how much energy is produced in a material as it is traversed by a given radiation. Table 5-1 illustrates several qualifying factors. The alpha particle referred to in Table 5-1 is the helium nucleus.

Table 5-1
Typical Qualifying Factors
[after Murray, 2001, pg. 213, Table 16.1]

x-rays, gamma rays	1
Thermal neutrons (0.025 electron volt)	2
High energy protons	10
Heavy ions, including alpha particles	20

Dose equivalent is expressed in sieverts (Sv) when the dose is expressed in Grays. R.L. Murray [2001, pg 214] wrote that a single, sudden dose of four sieverts can be fatal, while the typical annual exposure to natural and manmade radiation is 3.6 millisieverts (a millisievert is one thousandth of a sievert). Manmade radiation includes medical and dental radiation.

5.1.3 History of Nuclear Power

Physicist Leo Szilard conceived of a neutron chain reaction in 1934. Szilard knew that neutrons could interact with radioactive materials to create more neutrons. If the density of the radioactive material and the number of neutrons were large enough, a chain reaction could occur. A chain reaction occurs when neutrons released by a reaction cause more reactions, which release more neutrons that cause even more reactions. Szilard thought of two applications of the neutron chain reaction: a peaceful harnessing of the reaction for the production of consumable energy; and an uncontrolled release of energy (an explosion) for military purposes. Recognizing the potential significance of his concepts, Szilard patented them in an attempt to hinder widespread development of the military capabilities of the neutron chain reaction. This was the first attempt in history to control the proliferation of nuclear technology.

Italian physicist Enrico Fermi and his colleagues in Rome were the first to bombard radioactive material using low-energy ("slow") neutrons in 1935. The spatial extent of the nucleus is often expressed in terms of a unit called the Fermi, in honor of Enrico Fermi. One Fermi is roughly the diameter of a nucleus and is the range of the nuclear force that holds the nucleus together. The correct interpretation of Fermi's results as a nuclear fission process was provided in 1938 by Lise Meitner and Otto Frisch in Sweden, and Otto Hahn and Fritz Strassmann in Berlin. Hahn and Strassmann observed that neutrons colliding with uranium could cause the uranium to split into smaller elements. This process was called nuclear fission. The fission process produces smaller nuclei from the breakup of a larger nucleus, and can release energy and neutrons. If neutrons interact with the nuclei of fissionable material, it can cause the nuclei to split and produce more neutrons, which can then react with more fissionable material in a chain reaction. Fermi succeeded in operating the first sustained chain reaction on December 2, 1942 in a squash court at the University of Chicago.

The scientific discovery of radioactivity and nuclear fission did not occur in a peaceful society, but in a world threatened by the militaristic ambitions of Adolf Hitler and Nazi Germany. In 1939, Hitler's forces plunged the world into war. Many prominent German scientists fled to the United States and joined an Allied effort to develop the first nuclear weapons. Their effort, known as the Manhattan Project, culminated in the successful development of the atomic bomb utilizing nuclear fission. The first atomic bomb was detonated in the desert near Alamogordo, New Mexico in 1945.

By this time, Hitler's Germany was in ruins and there was no need to use the new weapon in Europe. Japan, however, was continuing the Axis fight against the Allied forces in the Pacific and did not seem willing to surrender without first being defeated on its homeland. Such a defeat, requiring an amphibious assault against the Japanese islands, would have cost many lives, both of combatants and Japanese non-combatants. United States President Harry Truman decided to use the new weapon.

The first atomic weapon to be used against an enemy in war was dropped by a United States airplane on the Japanese city of Hiroshima on August 6, 1945. Approximately 130,000 people were killed and 90% of the city was destroyed. When the Japanese government refused to surrender unconditionally, a second atomic bomb was dropped on the Japanese city of Nagasaki on August 9, 1945. The combined shocks of two nuclear attacks and a Soviet invasion of Manchuria on August 9, 1945 prompted the Japanese Emperor to accept the Allies' surrender terms on August 14, 1945 and to formally surrender aboard the USS Missouri (Figure 5-1) on September 2, 1945.

People throughout the world realized that nuclear weapons had significantly altered the potential consequences of an unlimited nuclear war. By the early 1950's, the Soviet Union had acquired the technology for building nuclear weapons and the nuclear arms race was on. As of 1972, the United States maintained a slight edge over the Soviet Union in strategic nuclear yield and almost four times the number of deliverable warheads. Yield is a measure of the explosive energy of a nuclear weapon. It is often expressed in megatons, where one megaton of explosive is equivalent to one million tons of TNT.

During the decade from 1972 to 1982, the size of the United States nuclear arsenal did not change significantly, but the quality of the weapons, particularly the delivery systems such as missiles, increased dramatically. By 1982, the Soviet Union had significantly closed the gap in the number of deliverable warheads and had surged ahead of the United States in nuclear yield. The Soviet Union and the United States had achieved a rough nuclear parity. The objective of this parity was a concept called "deterrence."

Deterrence is the concept that neither side will risk engaging in a nuclear war because both sides would suffer unacceptably large losses in life and property. This is the concept that underlies the doctrine of Mutual Assured Destruction: the societies of all participants in a nuclear war would be destroyed. A consequence of this policy is that the societies involved in such a pact must remain vulnerable to one another; otherwise the less vulnerable society has a strategic advantage that could threaten the more vulnerable society and lead to instability and war. The "star wars" concept advanced by United States President Ronald Reagan in the 1980's is a missile defense system that provided an alternative strategy to Mutual Assured Destruction, but threatened the global balance of nuclear power.

Figure 5-1. The USS Missouri berthed at Pearl Harbor, Hawaii (Fanchi, 2004)

The Soviet Union was unable to compete economically with the United States and dissolved into separate states in the late 1980's. Some of the states of the former Soviet Union, notably Russia and Ukraine, retain nuclear technology. Other nations around the world have developed nuclear technology for peaceful purposes, and possibly for military purposes as well.

Point to Ponder: Can we put the nuclear genie back in the bottle?
Some historians argue that nuclear weapons were not needed to end World War II and that the world would be better off if nuclear energy had never been developed. Today, nations with nuclear weapons arsenals must be concerned about the security of every weapon in its arsenal. Even one nuclear weapon can be a weapon of mass destruction. In addition, many nations are seeking to acquire nuclear technology, which increases the risk of nuclear weapons proliferation. Now that the world knows about nuclear energy, society must learn to govern its use. One forecast discussed in Chapter 14 is based on the premise that nuclear technology can be safely controlled and used as a long-term source of energy.

5.2 Nuclear Reactors

The primary commercial purpose of a nuclear reactor is to generate electric power from the energy released by a nuclear reaction. Nuclear reactors also provide power for ships such as submarines and aircraft carriers, and they serve as facilities for training and research. Our focus here is on the use of nuclear reactors to generate electricity. We then discuss global dependence on nuclear energy.

5.2.1 Nuclear Fission Reactors

The neutrons produced in fission reactions typically have energies ranging from 0.1 million electron volts (1.6×10^{-14} Joules) to 1 million electron volts (1.6×10^{-13} Joules). Neutrons with energies this high are called fast neutrons. Fast neutrons can lose kinetic energy in collisions with other materials in the reactor. Less energetic neutrons with kinetic energies on the order of the thermal energy of the reactor are called slow neutrons or thermal neutrons. Some nuclei tend to undergo a fission reaction after they capture a slow neutron. These nuclei are called fissile materials and are valuable fuels for nuclear fission reactors. Fission is induced in fissile materials by slow neutrons. Fission reactors that depend on the nuclear capture of fast neutrons exist, but are beyond the scope of this book.

Materials like light water (light water is ordinary water – H_2O), heavy water (deuterium oxide – D_2O), graphite and beryllium control the number of neutrons in the reactor. They are called moderating materials because they moderate, or control, the nuclear reaction. Moderating materials slow down fast neutrons by acquiring kinetic energy in collisions when a fast neutron collides with the more massive, slower moderating material. Some of the kinetic energy of the fast neutron is transferred to the moderating material and the speed of the fast neutron decreases.

Coolant materials transport the heat of fission away from the reactor core. Coolants include light water, carbon dioxide, helium, and liquid sodium. In some cases, moderating materials can function as coolants.

A pressurized water reactor heats coolant water to a high temperature and then sends it to a heat exchanger to produce steam. A boiling water reactor provides steam directly from water in contact with the fissile materials. The steam from each of the reactors turns a turbine to generate electricity. A pressurized water reactor is sketched in Figure 5-2. A nuclear reactor is housed in a containment building (Figure 5-3). The containment building, which is typically a dome, provides protection from internal leaks as well as external dangers, such as an airplane crash.

Nuclear fission reactor design and operation depends on the ratio of neutrons in succeeding generations of decay products. The ratio of the number of neutrons in one generation to the number of neutrons in the preceding generation is called the neutron multiplication factor. For

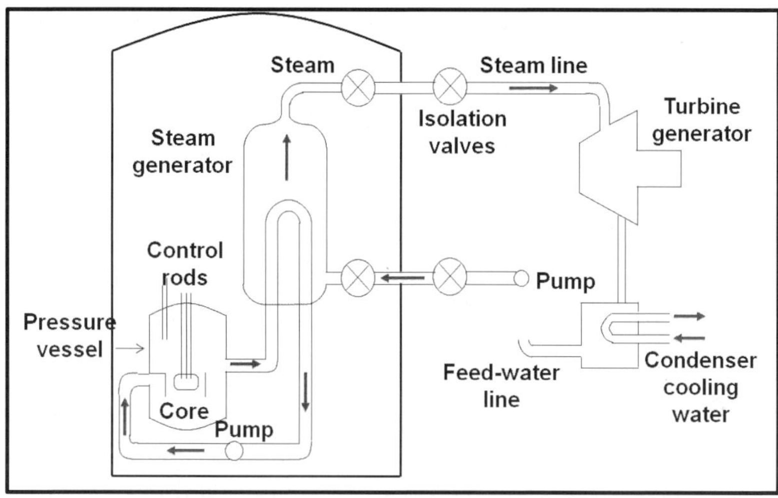

Figure 5-2. Schematic of a Pressurized Water Reactor [after Cassedy and Grossman, 1998, page 177; and U.S. Department of Energy Report Energy Technologies and the Environment, Report Number DOE/EP0026]

Figure 5-3. Containment Building at San Onofre Nuclear Generating Station, California [Fanchi, 2002]

example, if 100 neutrons produced in one generation of nuclear reactions interact with fissile material and produce 110 neutrons in the next generation, the neutron multiplication factor is 1.1 (that is, 110/100). The neutron multiplication factor is a positive quantity that depends on several factors [Lilley, 2001, Chapter 10].

If the multiplication factor for a reactor is less than one, the number of fission reactions will diminish from one generation to the next and the chain reaction will die out. The reactor level is considered subcritical. A critical mass of fissile material is present in the reaction chamber when there are enough neutrons being produced by nuclear reactions to balance the loss of neutrons. When the reaction is at the critical level, the multiplication factor for a reactor is equal to one and the chain reaction can be sustained. If the multiplication factor is greater than one, the number of neutrons produced by nuclear reactions exceeds the loss of neutrons and the number of fission reactions will increase from one generation to the next. The reactor level is considered supercritical. The chain reaction in a supercritical reactor will accelerate and, if it proceeds too fast, can have explosive consequences. The nuclear fission bomb is a supercritical reactor.

Products of fission reactions include nuclei that emit biologically harmful radiation. The nuclei are contained in spent fuel rods. Some of the nuclei, such as radioactive isotopes of cesium and iodine, can be radioactive for thousands of years. This material is considered radioactive waste and poses a significant environmental problem. The disposal of radioactive waste is discussed further in Section 5.4.

5.2.2 Availability of Nuclear Fuel

The most abundant fuel for nuclear fission is uranium. Uranium exists in the crust of the earth as the mineral uraninite. Uraninite is commonly called pitchblende and is a uranium oxide (U_3O_8). It is found in veins in granites and other igneous rocks. It is possible to find uranium in sedimentary rocks. In this case, scientists believe that uraninite was precipitated in sedimentary rocks after being transported from igneous rocks by the flow of water containing dissolved uraninite.

Uranium is obtained by mining for the mineral uraninite. Mining methods include underground mining, open pit mining, and in situ leaching. Leaching is a process of selectively extracting a metal by a chemical reaction that creates a water-soluble molecule that can be transported to a recovery site. The isotope of uranium that undergoes spontaneous fission (uranium-235) is approximately 0.7% of naturally occurring uranium ore. Uranium must be separated from mined ore and then enriched for use in nuclear fission reactors. The enrichment process is designed to purify uranium-235 and is often the main obstacle faced by countries seeking nuclear fission technology for use in energy generation or weapons.

Other fuels that can be used in the fission process include the fission products plutonium-239 and thorium-232. Specialized reactors called breeder reactors are designed to operate with fuels other than uranium. A breeder reactor is a nuclear fission reactor that produces more fissile material than it consumes.

The amount of uranium that can be recovered from the earth is called uranium reserves. Table 5-2 presents estimates of uranium reserves in the United States for the most common mining methods. Estimates of uranium reserves have been made and have many of the same uncertainties associated with estimates of fossil fuel reserves. Factors that are not well known include the distribution and extent of uranium deposits and the price people are willing to pay to recover the resource.

Uranium is considered a non-renewable resource because it exists as a finite volume within the earth. A comparison of the two columns of uranium oxide reserves in Table 5-2 shows that the amount of uranium oxide reserves increases as price per pound increases. The supply of uranium increases if you are willing to pay for it.

Two more sources of uranium that could be used for generating electrical power are sea water and nuclear weapons. Reduction in nuclear weapons stockpiles and dismantling nuclear warheads would provide some enriched uranium. A larger source of uranium is seawater. Table 5-3 shows typical concentrations of uranium in parts per million (ppm) from the World Nuclear Association website. The concentration of uranium in sea water is small, but the volume of sea water is huge. Technology for extracting uranium from sea water exists on the laboratory scale, but is currently too expensive for large-scale commercial application.

Table 5-2
United States Uranium Reserves by Mining Method and Cost, December 2003 Estimate [US EIA, http://www.eia.doe.gov/cneaf/nuclear/page/reserves/uresmine.html, accessed December 21, 2009]

Mining Method	US$30 per pound	US$50 per pound
	Uranium Oxide (U_3O_8) million pounds	Uranium Oxide (U_3O_8) million pounds
Underground	138	464
Open Pit	29	257
In Situ Leaching	98	165
Low grade ore, etc.	< 1	4
TOTAL	265	890

Table 5-3
Concentration of Uranium [WNA, September 2009]

Very high-grade ore (Canada) - 20% U	200,000 ppm U
High-grade ore - 2% U,	20,000 ppm U
Low-grade ore - 0.1% U,	1,000 ppm U
Very low-grade ore (Namibia) - 0.01% U	100 ppm U
Granite	4-5 ppm U
Sedimentary rock	2 ppm U
Earth's continental crust (av)	2.8 ppm U
Seawater	0.003 ppm U

The geological distribution of uranium resources is illustrated by showing the ten countries with the largest known recoverable uranium resources in the world as of January 1, 2007. These countries are listed in Table 5-4. The list is based on information from the Organization for Economic Cooperation and Development (OECD) and a recovery cost of US$60 per pound. Resources listed in Table 5-4 do not include uranium in sea water.

Table 5-4
Distribution of Uranium Resources as of 1/1/2007
[WNA, September 2009]

Country	Million Pounds	Percent of World
Australia	2740	23%
Kazakhstan	1801	15%
Russia	1204	10%
South Africa	959	8%
Canada	933	8%
United States	754	6%
Brazil	613	5%
Namibia	606	5%
Niger	604	5%
Ukraine	441	4%
World	12057	

5.2.3 Nuclear Fusion Reactors

One of the appealing features of nuclear fusion is the relative abundance of hydrogen and its isotopes compared to fissile materials. The idea behind nuclear fusion is quite simple: fuse two molecules together and release large amounts of energy in the process. Examples of fusion reactions include collisions between protons, deuterons (deuterium nuclei), and tritons (tritium nuclei). Protons are readily available as hydrogen nuclei. Deuterium is also readily available. Ordinary water contains approximately 0.015 mole % deuterium [Murray, 2001, page 77], thus one atom of deuterium is present in ordinary water for every 6700 atoms of hydrogen.

The fusion reaction can occur only when the atoms of the reactants are heated to a temperature high enough to strip away all of the atomic electrons and allow the bare nuclei to fuse. The state of matter containing bare nuclei and free electrons at high temperatures is called plasma. Plasma is an ionized gas. The temperatures needed to create plasma and allow nuclear fusion are too high to be contained by conventional building materials. Two methods of confining plasma for nuclear fusion are being considered.

We first consider magnetic confinement, which is a confinement method that relies on magnetic fields to confine the plasma. The magnetic confinement reactor is called a tokamak reactor. Tokamak reactors are toroidal (donut shaped) magnetic bottles that contain the plasma that is to be used in the fusion reaction. Plasma is a gas of ionized particles. Two magnetic fields confine the plasma in a tokomak: one is provided by cylindrical magnets that create a toroidal magnetic field; and the other is a poloidal magnetic field that is created by the plasma current. Combining these two fields creates a helical field that confines the plasma. Existing tokamak reactors inject deuterium and tritium into the vacuum core of the reactor at very high energies. Inside the reactor, the deuterium and tritium isotopes lose their electrons in the high energy environment and become plasmas. The plasmas are confined by strong magnetic fields until fusion occurs.

The current fusion power record is presently held by the Joint European Torus (JET) in Oxfordshire, United Kingdom. JET is a magnetic confinement fusion reactor based on the tokamak concept. The record is 16 MW and was sustained for one to two seconds [Feder, 2009].

The second confinement method is inertial confinement. Inertial confinement uses pulsed energy sources such as lasers to concentrate energy onto a small pellet of fusible material, such as a frozen mixture of deuterium and tritium. The pulse compresses and heats the pellet to ignition temperatures.

Point to Ponder: Why is nuclear fusion so desirable?
Nuclear fusion is considered an environmentally clean source of energy. Unlike nuclear fission, nuclear fusion does not generate waste products that are lethal for thousands of years. The material needed for nuclear fusion is much more plentiful than the known supply of fissionable materials, such as uranium.

The fusion reactor concept can be combined with the fission process to create a nuclear reactor system that is a fusion-fission hybrid. A fusion reactor would serve as a neutron source for a fission reactor. The hybrid concept could produce power, burn nuclear waste, and produce fissile material. The combination of fusion and fission technologies would, however, be more complicated and expensive than working with either a fusion system or a fission system.

5.2.4 ITER: A Prototype Nuclear Fusion Reactor

Many scientists expect nuclear energy to be provided by nuclear fusion sometime during the 21st century. Attempts to harness and commercialize fusion energy have so far been unsuccessful because of the technical difficulties involved in igniting and controlling a fusion reaction. Nuclear fusion must be controlled before it can be commercialized. A prototype nuclear fusion reactor called the International Thermonuclear Experimental Reactor (ITER) is presently under construction in Cadarache, France [ITER, 2009].

The ITER project is sponsored by China, the European Union, India, Japan, Korea, Russia, and the United States. The acronym ITER is Latin for "the way" and was selected in part to signify the way to new energy. The ITER reactor is a commercial-scale tokamak reactor that will use deuterium and tritium as nuclear fuel. ITER is expected to consume 50

MW of power to produce 500 MW of nuclear fusion power. The reactor will be the precursor of a demonstration nuclear fusion power plant.

> **Point to Ponder: When will nuclear fusion be commercially available?**
> A commercial nuclear fusion reactor has not yet been constructed. Nuclear fusion reactors are still in the development stage and are not expected to provide commercial power for a generation. The prototype nuclear fusion reactor ITER is scheduled for completion in 2018, and fusion energy is expected to contribute significantly to the energy mix by the end of the 21st century.

5.3 Global Dependence on Nuclear Power

The first commercial nuclear power plant was built on the Ohio River at Shippingport, Pennsylvania, a city about 25 miles from Pittsburgh. It began operation in 1957 and generated 60 megawatt of electric power output [Murray, 2001, page 202]. Today, nuclear power plants generate a significant percentage of electricity in some countries.

Table 5-5 lists the top ten producers of electric power from nuclear energy and their percentage of the world's total electric power generation from nuclear energy for the year 2006. The total electric power generated from nuclear energy in the world was 2,660 billion kilowatt-hours in 2006. Total electric power generated in the world that same year was 18,014 billion kilowatt-hours. Consequently, electric power generated from nuclear energy provided approximately 14.8% of the electric power generated in the world in 2006.

Some of the nations that use nuclear energy to produce electricity also have nuclear weapons, but some of the nations do not. For example, the United States, Russia, and France have nuclear weapons while Japan and Germany do not. The decision to enrich uranium so that it can be used in a nuclear weapon depends on factors such as the history of the nation and its politics. The United States, for example, developed nuclear weapons in an effort to quickly end World War II. Japan and Germany, on the other hand, were defeated in World War II and their militaries were subject to restrictions imposed by the victors, notably the United States, Russia, Great Britain, and France.

Table 5-5
Top Ten Producers of Electric Power from Nuclear Energy in 2006
[US EIA, EIA website, accessed December 22, 2009]

Country	Electric Power from Nuclear Energy (billion kWh)	Percentage of World Electric Power from Nuclear Energy (World = 2,660 billion kWh)
United States	787.2	29.6
France	427.7	16.1
Japan	288.3	10.8
Germany	144.3	6.0
Russia	141.3	5.4
Korea, South	93.1	5.3
Canada	84.8	3.5
Ukraine	71.7	3.2
United Kingdom	63.6	2.7
Sweden	57.1	2.4

Some countries are highly dependent on nuclear energy. Table 5-6 shows the percentage of electric power generated from nuclear energy compared to total electric power generation for the top ten producers of electric power from nuclear energy for the year 2006. Nuclear energy was used to generate most of the electric power generated in France. Ukraine and Sweden use nuclear power to generate approximately 40% of their electricity.

Table 5-6
Dependence of Nations on Nuclear Energy in 2006
[US EIA, EIA website, accessed December 22, 2009]

Country	Electric Power from Nuclear Energy (billion kWh)	Total Electricity Generation (billion kWh)	Nuclear Share (% National Total)
United States	787.2	4071	19.3
France	427.7	542	78.9
Japan	288.3	1032	27.9
Germany	144.3	595	24.3
Russia	141.3	941	15.0
Korea, South	93.1	380	24.5
Canada	84.8	595	14.3
Ukraine	71.7	182	39.4
United Kingdom	63.6	372	17.1
Sweden	57.1	139	41.1

5.4 Nuclear Energy and the Environment

The selection of an environmentally compatible primary fuel is not a trivial problem. Two energy sources of special concern for the 21st century energy mix are nuclear fission and nuclear fusion. We consider their environmental impact in more detail here.

5.4.1 Nuclear Fission

Compared to fossil fuel driven power plants, nuclear fission plants require a relatively small mass of resource to fuel the nuclear plant for an extended period of time. Nuclear fission plants rely on a non-renewable resource: uranium-235. The Earth's inventory of uranium-235 will eventually be exhausted. Breeder reactors use the chain reaction that occurs in the reactor control rods to produce more fissionable material (specifically plutonium-239).

One of the main concerns of nuclear fission technology is to find a socially and environmentally acceptable means of disposing of fuel rods containing highly radioactive waste. The issue of waste is where most of the debate about nuclear energy is focused. The waste generated by nuclear fission plants emits biologically lethal radiation and can contaminate the site where it is stored for thousands of years. On the other hand, environmentally compatible disposal options are being developed. One disposal option is to store spent nuclear fuel in geologically stable environments.

A nuclear fission power plant can contaminate air, water, the ground, and the biosphere. Air can be contaminated by the release of radioactive vapors and gases through water vapor from the cooling towers, gas and steam from the air ejectors, ventilation exhausts, and gases removed from systems having radioactive fluids and gases. The radiation released into the air can return to the earth as radiated rain, which is the analog to acid rain generated by the burning of fossil fuels. Water may be contaminated when radioactive materials leak into coolant water. The contaminated water can damage the environment if it is released into nearby bodies of water such as streams or the ocean. Water and soil contamination can occur when radioactive waste leaks from storage containers and seeps into underground aquifers. The biosphere (people, plants, and animals) is affected by exposure to radioactive materials in the environment. The effect of exposure is cumulative and can cause the immune system of an organism to degrade.

5.4.2 Radioactive Waste

Nuclear power plants generate radioactive wastes that require long-term storage. Some of the end products of nuclear fission are highly radioactive and have a half-life measured in thousands of years. They must be disposed in a way that offers long-term security. One solution is to place used uranium rods containing plutonium and other dangerously radioactive compounds in an isolated location that is geologically stable. The location should not be significantly affected by low levels of radiation.

Figure 5-4. Yucca Mountain

A possible location for storing nuclear waste in the United States is Yucca Mountain, Nevada. Figure 5-4 shows an aerial view of Yucca Mountain and Figure 5-5 shows a train inside the Yucca Mountain tunnel. Yucca Mountain was chosen for long-term nuclear waste storage by the Federal government, but local public resistance has delayed the waste disposal project.

Figure 5-5. Train inside Yucca Mountain Tunnel

A Waste Isolation Pilot Plant (WIPP) in the Chihuahuan Desert near Carlsbad, New Mexico began disposal operations in 1999. It was built by the United States Department of Energy and carved out of the Permian Salt Formation nearly a half mile below the surface. The WIPP site is used to dispose defense-related radioactive waste.

Point to Ponder: Can nuclear waste be stored safely?
Natural nuclear fission reactors were discovered by French physicist Francis Perrin at the Oklo uranium mine in the equatorial West African republic of Gabon in 1972 [WNA, March 2001; DoE Oklo, 2009]. Scientists believe that the ore deposit, which contained about 3% uranium-235, began a self-sustaining nuclear fission chain reaction millions of years ago. The natural reactors created radioactive waste that included fission products and elements found today in used nuclear fuel. The World Nuclear Association said that "The Oklo chain reaction occurred intermittently for more than 500,000 years. Despite its location

in a wet, tropical climate, Oklo's uranium deposit and high-level waste has remained securely locked in this natural repository for the past 2000 million years. Many of the waste products stayed where they were created or moved only a few centimeters before decaying into harmless products." The natural nuclear fission reactors are natural analogues that demonstrate that it is possible to dispose nuclear waste in a secure and safe manner for an indefinite period of time.

5.4.3 Containment Failure

Radioactive materials can be released from nuclear reactors if there is a failure of the containment system. Containment can be achieved by the reactor vessel and by the containment dome enclosing the reactor vessel. The two most publicized nuclear power plant incidents were containment failures. They occurred at Three Mile Island, Pennsylvania in 1979 and at Chernobyl, Ukraine in 1986.

The Three Mile Island power plant was a pressurized water reactor that went into operation in 1978 and produced approximately a gigawatt of energy. The containment failure at Three Mile Island occurred on March 29, 1979. It began when coolant feedwater pumps stopped and temperature in the reactor vessel began to rise. An increase in pressure accompanied the increasing temperature and caused a pressure relief valve to open. The reactor shut down automatically. Steam from the reactor flowed through the open relief valve into the containment dome. The valve failed to shut at a pre-specified pressure and vaporized coolant water continued to flow out of the reactor vessel through the open valve. The water-steam mixture flowing through the coolant pumps caused the pumps to shake violently. Plant operators did not realize they were losing coolant and decided to shut off the shaking pumps. A large volume of steam formed in the reactor vessel and the overheated nuclear fuel melted the metal tubes holding the nuclear fuel pellets. The exposed pellets reacted with water and steam to form a hydrogen gas bubble. Some of the hydrogen escaped into the containment dome. The containment dome did not fail; it contained the hydrogen gas bubble and pressure fluctuations. The operators were eventually able to disperse the hydrogen bubble and regain control of the reactor.

The Chernobyl containment failure occurred in a boiling water reactor that produced a gigawatt of power and had a graphite moderator. Operators at the plant were using one of the reactors, Unit 4, to conduct an experiment. They were testing the ability of the plant to provide electrical power as the reactor was shut down. To obtain measurements, the plant operators turned off some safety systems, such as the emergency cooling system, in violation of safety rules. The operators then withdrew the reactor control rods and shut off the generator that provided power to the cooling water pumps. Without coolant, the reactor overheated. Steam explosions exposed the reactor core and fires started. The Chernobyl reactors were not encased in massive containment structures that are common elsewhere in the world. When the explosions exposed the core, radioactive materials were released into the environment and a pool of radioactive lava burned through the reactor floor. The Chernobyl accident was attributed to design flaws and human error.

Except for the Chernobyl incident, no deaths have been attributed to the operation of commercial nuclear reactors. M.W. Carbon [1997, Chapter 5] reported that the known death toll at Chernobyl was less than fifty people. R.A. Ristinen and J.J. Kraushaar [1999, Section 6.9] reported an estimate that approximately 47,000 people in Europe and Asia will die prematurely from cancer because they were exposed to radioactivity from Chernobyl. The disparity in the death toll associated with the Chernobyl incident illustrates the range of conclusions that can be drawn by different people with different perspectives.

5.4.4 Nuclear Fallout

Nuclear fallout is the deposition of radioactive dust and debris that was carried into the atmosphere by the detonation of a nuclear weapon. Nuclear fallout can also be generated by the release of radioactive material into the atmosphere. Radioactive material can be released by a variety of events such as reactor containment failure or detonation of a dirty bomb made by combining conventional explosives and radioactive waste. The pattern of deposition of the fallout depends on climatic conditions such as ambient air temperature and pressure, and wind conditions.

5.4.5 Nuclear Winter

Nuclear winter is the phrase used to describe a decline of air temperature resulting from an increase of particulates in the atmosphere following the detonation of many nuclear weapons. The temperature decline would generate wintry conditions, hence the phrase "nuclear winter." A decline in air temperature may have occurred during extinction events associated with meteors striking the earth. One example of such an extinction event was the disappearance of the dinosaurs approximately 65 million years ago.

In the nuclear winter scenario, an increase of particulates in the atmosphere increases the earth's albedo, which quantifies the reflection of incident sunlight away from Earth. The albedo increases when the amount of incident sunlight reflected by the atmosphere increases. Proponents of the nuclear winter scenario argue that an increase in the reflection of incident solar radiation by the atmosphere can cause a reduction of air temperature. Critics of the nuclear winter scenario argue that the greenhouse effect may tend to increase the surface temperature of the earth. It is now believed that the decline of air temperature would not be as severe as originally thought.

5.4.6 Nuclear Fusion

The isotopes of hydrogen needed for nuclear fusion are abundant. The major component of fusion, deuterium, can be extracted from water, which is available around the world. Tritium, another hydrogen isotope, is readily available in lithium deposits that can be found on land and in seawater. Unlike fossil fuel driven power plants, nuclear fusion does not emit air pollution.

Nuclear fusion reactors are considered much safer than nuclear fission reactors. The amounts of deuterium and tritium used in the fusion reaction are so small that the instantaneous release of a large amount of energy during an accident is highly unlikely. The fusion reaction can be shut down in the event of a malfunction with relative ease. A small release of radioactivity in the form of neutrons produced by the fusion reaction may occur, but the danger level is much less than that of a fission reactor.

The main problem with nuclear fusion energy is that the technology is still under development: commercially and technically viable nuclear fusion reactor technology does not yet exist. A panel working for the United States Department of Energy has suggested that nuclear fusion could be providing energy to produce electricity by the middle of the 21st century if adequate support is provided to develop nuclear fusion technology [Dawson, 2002]. The ITER project discussed previously is an example of a nuclear fusion concept that has potential for future commercialization.

Point to Ponder: Is nuclear power socially acceptable?
Nuclear power has been socially acceptable in some political jurisdictions around the world. Some European states, notably France, are reliant on nuclear fission power. The acceptability of nuclear power depends on such factors as finding an environmentally acceptable solution for the storage of nuclear wastes and the cost of energy using non-nuclear energy sources. As the price of non-nuclear energy to the consumer increases, nuclear energy may become more appealing. The environmental concern over the geologic storage of radioactive waste must be weighed against the environmental concern over the emission and accumulation of greenhouse gases in the atmosphere because of the continuing reliance on fossil fuels.

5.5 ACTIVITIES

True-False

Specify if each of the following statements is True (T) or False (F).

1. Nuclear fission is the splitting of a large nucleus into smaller nuclei.
2. The sun provides a finite source of energy using nuclear fusion.
3. Leo Szilard was the first person to attempt to prevent nuclear technology proliferation.
4. Water can be used as both a moderating and cooling material in nuclear fission reactors.
5. A pressurized water reactor provides steam directly from the reactor to turn a turbine.
6. A breeder reactor consumes more fissile material than it produces.
7. Nuclear fission is a non-renewable energy source.
8. The fuel for nuclear fusion is cleaner and more plentiful than the fuel for nuclear fission.
9. Isotopes are nuclei with the same atomic number but different numbers of neutrons.
10. Energy from the sun is the result of nuclear fusion.

Questions

1. List 5 nations that produce electric energy from nuclear energy. Which of these countries use commercial nuclear fusion power plants?
2. Match the site to the events by writing the letter of the event in the column to the left of the site.

	Site		Event
	Alamogordo	A	Nuclear power plant accident with many deaths
	Chernobyl	B	First city bombed by a nuclear weapon in World War II
	Yucca Mountain	C	First nuclear weapon detonation
	Hiroshima	D	Nuclear waste storage

3. What is nuclear winter?
4. What is Mutual Assured Destruction?

5. How do you control the number of neutrons in a nuclear fission reactor?
6. What is the purpose of a coolant material in a nuclear fission reactor?
7. What is the purpose of a containment building?
8. What is the difference between nuclear fission and nuclear fusion?
9. List the two methods of confining plasma used in nuclear fusion reactors.
10. What is a tokomak reactor and what does it do?

CHAPTER 6

RENEWABLE ENERGY – SOLAR ENERGY

In Chapter 1, we discussed the history of fuel use and mentioned that major technological advances are often the catalyst for change in the global energy mix. These advances include such developments as the steam engine, discovery of oil's use as an illuminant, Edwin L. Drake's first successful oil well, and more recently the discovery of atomic power. Today, the catalyst for change is different, but the need for new technology is still a limiting factor. Today's energy mix is being driven by two primary factors: the rising cost and diminishing reserves of fossil fuels, and growing concerns about the damage carbon emissions are causing to the environment. These factors have motivated interest in renewable energy.

Renewable energy is energy obtained from sources at a rate that is less than or equal to the rate at which the source is replenished. Energy consumed by the United States in 2008 included 7.4% renewable energy (Appendix B). World energy consumption included almost 9% renewable energy in 2006 (Appendix C). Solar energy provided approximately 1% of the renewable energy produced in the United States in 2008 (Appendix B), and the cost of electricity from solar energy is 5 to 10 times more than electricity from fossil fuels [Crabtree and Lewis, 2007]. For comparison, about 7% of the renewable energy consumed by the United States in 2008 was wind, and 34% was hydroelectric. Wind energy and hydroelectric power are discussed further in Chapters 7 and 8, respectively.

The production of electrical power from renewable energy sources does not emit as much pollution as power production from fossil fuels. As for cost, most renewable energy will become less expensive over time as the infrastructure for harvesting and delivering energy matures. The most likely sources of renewable energy that are expected to contribute to the future energy mix are solar energy, wind energy, energy from water, and biofuels. Many of these are already becoming significant contributors to the global energy supply today.

The term "renewable" is somewhat of a misnomer when applied to solar energy. Solar energy is provided by the sun. Since the remaining lifetime of the sun is expected to be a few billion years, many people consider solar energy an inexhaustible supply of energy. In fact, solar energy from the sun is finite, but should be available for use by many generations of people. Solar energy is therefore considered renewable. Energy sources that are associated with solar energy, such as wind and biomass, are also considered renewable.

Solar radiation may be converted to other forms of energy by several conversion processes. Thermal conversion relies on the absorption of solar energy to heat a cool surface. Biological conversion of solar energy relies on photosynthesis. Photovoltaic conversion generates electrical power by the generation of an electrical current as a result of a quantum mechanical process. Wind power and ocean energy conversion rely on atmospheric pressure gradients and oceanic temperature gradients to generate electrical power. In this chapter we focus on thermal conversion and photovoltaic conversion of solar energy in three forms: passive, active, and electric. Passive and active solar energy are generally used for space conditioning, such as heating and cooling, while solar electric energy is used to generate electrical power. Each of these forms is discussed below.

6.1 SOURCE OF SOLAR ENERGY

Solar energy is energy emitted by a star. Figure 6-1 shows the anatomy of a star. Energy emitted by a star is generated by nuclear fusion. The fusion process occurs in the core, or center, of the star. Energy released by the fusion process propagates away from the core by radiating from one atom to another in the radiation zone of the star. As the energy moves away from the core and passes through the radiation zone, it reaches the part of the star where the energy continues its journey towards the surface of the star as heat associated with thermal gradients. This part of the star is called the convection zone. The surface of the star, called the photosphere, emits light in the visible part of the electromagnetic spectrum. The star is engulfed in a stellar atmosphere called the chromosphere. The chromosphere is a layer of hot gases surrounding the photosphere.

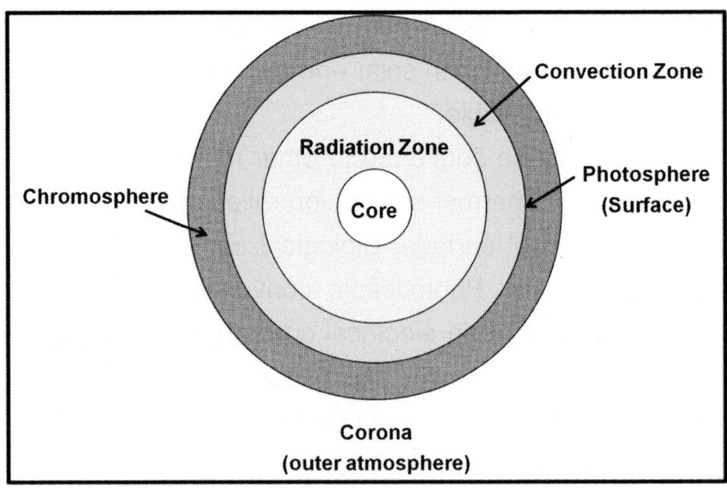

Figure 6-1. Anatomy of a Star

The amount of sunlight that reaches the earth depends on the motion of the earth around the sun. The orbit of the earth around the sun lies in a geometrical plane called the orbital plane. The ecliptic plane is the plane of the orbit that intersects the sun. The ecliptic plane and the orbital plane are shown in Figure 6-2. The line of intersection between the orbital plane and the ecliptic plane is the line of nodes. Most planetary orbits, including the earth's orbit, lie in the ecliptic plane.

The luminosity of a star is the total energy radiated per second by the star. The amount of radiation from the sun that reaches the earth's atmosphere is called the solar constant. The solar constant varies with time because the earth follows an elliptical orbit around the sun and the axis of rotation of the earth is inclined relative to the plane of the earth's orbit. The distance between points on the surface of the earth and the sun varies throughout the year.

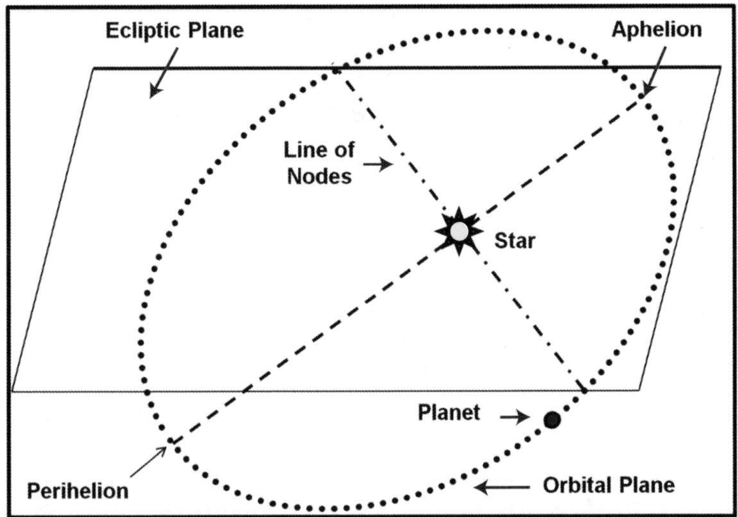

Figure 6-2. Planetary Orbit and the Ecliptic Plane

The amount of solar radiation that reaches the surface of the earth depends on the factors illustrated in Figure 6-3. The flux of solar radiation incident on a surface placed at the edge of the earth's atmosphere depends on the time of day and year, and the geographical location of the surface. Some incident solar radiation is reflected by the earth's atmosphere. The fraction of solar radiation that is reflected back into space by the earth-atmosphere system is called the albedo.

Approximately thirty-five percent of the light from the sun does not reach the surface of the earth. This is due to clouds (20%), atmospheric particles (10%), and reflection by the earth's surface (5%). The solar flux that enters the atmosphere is reduced by the albedo. Once in the atmosphere, solar radiation can be absorbed in the atmosphere or scattered away from the earth's surface by atmospheric particulates such as air, water vapor, dust particles, and aerosols. Some of the light that is scattered by the atmosphere eventually reaches the surface of the earth as diffused light. Solar radiation that reaches the earth's surface from the disk of the sun is called direct solar radiation if it has experienced negligible change in its original direction of propagation.

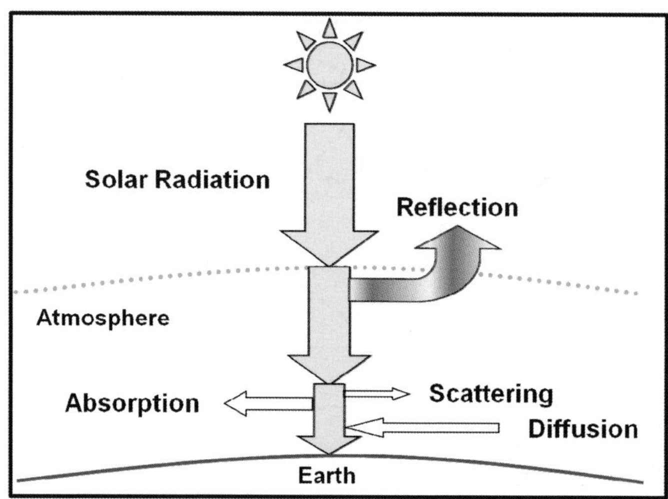

Figure 6-3. Solar Radiation and the Earth-Atmosphere System

Point to Ponder: Can we collect light from outer space?
We have seen that the atmosphere reduces the amount of sunlight reaching the surface of the earth. It is conceivable to collect sunlight above the surface of the earth and transmit it as an intense beam through the atmosphere. The goal would be to reduce the effective albedo of the atmosphere. Obviously, this type of space-age technology will be expensive. There would also be environmental and safety concerns for anything crossing the path of the beam.

In November 2009, Japan announced plans for the multi-billion dollar Space Solar Power System (SSPS) project which is designed to gather solar power from space [Poupee, 2009]. The project is a massive renewable energy project by one of the world's most-developed countries with the least natural energy resources. Japan has been an oil importer for most of its modern history because it lacks oil reserves. The lack of energy resources and a desire to be energy independent has motivated Japanese leaders to develop renewable energy. The SSPS will collect solar energy using solar photovoltaic technology, described below, and plans to beam the energy to the ground using a collection of lasers or microwaves. Microwaves are electromagnetic waves that can pass through atmospheric hindrances such as clouds, rain, dust, and smog.

6.2 Energy Conversion Efficiency and Intermittency

Efficiency is a concept that facilitates the comparison of different energy sources. Estimates of efficiency can help us quantify the total amount of energy that is needed to provide a smaller amount of energy in a useful form. Here we consider energy conversion efficiency, and intermittency.

Energy conversion efficiency is the ratio of the useful energy output by an energy conversion system to the energy input to the system. For example, if a power plant consumes 100 J of energy to generate 50 J of energy, the energy conversion efficiency is 50%. The energy that does not appear as useful output energy is used to operate the system or it is lost as waste energy. Waste energy often appears as heat. The energy balance must include useful energy output as well as work and energy loss.

Another factor that must be considered when comparing different energy systems is the availability of the system. Fossil fuels and nuclear energy can provide energy on demand. Renewable energy sources such as solar and wind are considered intermittent energy sources because they are not always available. Solar energy relies on access to sunlight, which is not available at night and is reduced by atmospheric conditions such as dust, clouds and smog. Wind varies in magnitude and direction depending on weather conditions. One way to mitigate intermittency is to use energy storage systems that harvest energy when the energy source is available, and then release the stored energy when it is needed.

6.3 Passive Solar Energy

Passive solar energy technology integrates building design with environmental factors that enable the capture or exclusion of solar energy. Mechanical devices are not used in passive solar energy applications. We illustrate passive solar energy technology by considering two simple but important examples: the roof overhang and thermal insulation.

6.3.1 Roof Overhang

Sunlight that strikes the surface of an object and causes an increase in temperature of the object is an example of direct solar heat. Direct solar heating can cause an increase in temperature of the interior of buildings with windows. The windows that allow in the most sunlight are facing south in the northern hemisphere and facing north in the southern hemisphere. Figure 6-4 illustrates two seasonal cases. The figure shows that the maximum height of the sun in the sky varies from season to season because of the angle of inclination of the earth's axis of rotation relative to the ecliptic plane. The Earth's axis of rotation is tilted 23.5° from a line that is perpendicular to the ecliptic plane.

One way to control direct solar heating of a building with windows is to build a roof overhang. The roof overhang is used to control the amount of sunlight entering the windows. Figure 6-4 illustrates a roof overhang. The longest shadow from a roof overhang is obtained during the summer, while the shortest shadow from a roof overhang is obtained during the winter. The shadow length can be changed by changing the length of the roof overhang.

Passive solar cooling is achieved when the roof overhang casts a shadow over the windows facing the sun. In this case, the roof overhang is designed to exclude sunlight, and its associated energy, from the interior of the building. Alternatively, the windows may be tinted with a material that reduces the amount of sunlight entering the building. Another way to achieve passive solar cooling is to combine shading with natural ventilation.

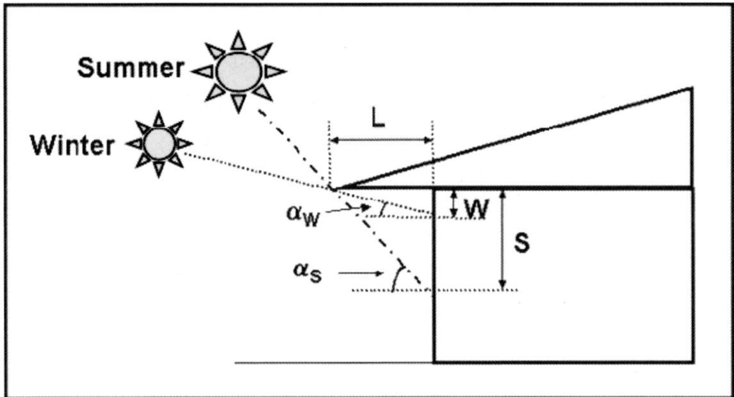

Figure 6-4. Roof Overhang

Passive solar heating is the capture and conversion of solar energy into thermal energy. The technology for passive solar heating can be as simple as using an outdoor clothesline to dry laundry or designing a building to capture sunlight during the winter. In the latter case, the building should be oriented to collect sunlight during cooler periods. Sunlight may enter the building through properly positioned windows that are not shaded by a roof overhang, or through skylights. The sunlight can heat the interior of the building and it can provide natural light. The use of sunlight for lighting purposes is called daylighting. An open floor plan in the building interior maximizes the effect of daylighting and can substantially reduce lighting costs.

6.3.2 Thermal Conductivity and Insulation

Solar energy may be excluded from the interior of a structure by building walls that have good thermal insulation. The quality of thermal insulation for a wall with the geometry shown in Figure 6-5 can be expressed in terms of thermal conductivity and thermal resistance.

The rate of heat flow through the insulated wall shown in Figure 6-5 depends on wall thickness, the cross-sectional area perpendicular to the direction of heat flow, and the temperature difference between the exposed and inside faces of the wall. The rate of heat flow through the insulated wall depends on a property of the wall called thermal conductivity. Thermal conductivity is a measure of heat flow through a material. Metals have relatively high thermal conductivities.

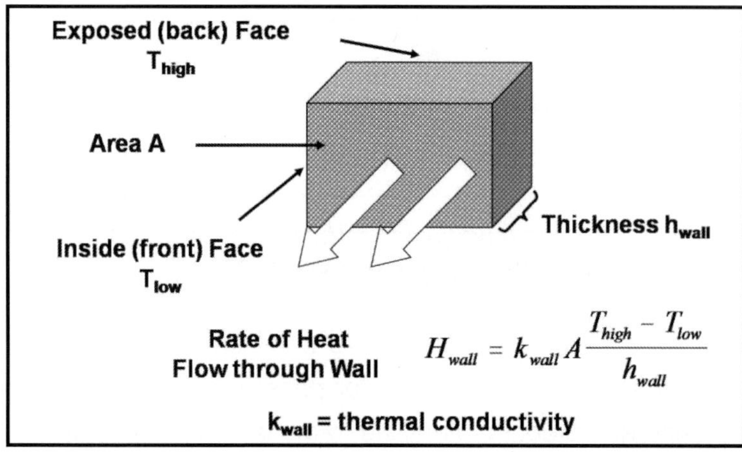

Figure 6-5. Thermal Conductivity of an Insulated Wall

We have considered thermal insulation as a passive solar technology example. Thermal insulation is also an energy conservation technology. Thermal insulation in walls can keep heat out of a structure during the summer and keep heat in during the winter. Consequently, thermal insulation can reduce the demand for energy to cool a space during the summer and heat a space during the winter. This reduces the demand for energy and makes it possible to conserve, or delay, the use of available energy.

6.4 Active Solar Energy

Active solar energy refers to the design and construction of systems that collect and convert solar energy into other forms of energy such as heat and electrical energy. Active solar energy technologies are typically energy conversion systems that are used to collect and concentrate solar energy, and convert to a more useful form of energy. One way to compare different energy conversion processes is to calculate their energy conversion efficiencies. We discuss solar heat collectors in this section, and solar power plants in the next section as illustrations of active solar energy technology.

6.4.1 Solar Heat Collectors

Solar heat collectors capture sunlight and transform radiant energy into heat energy. Figure 6-6 is a diagram of a solar heat collector. Sunlight enters the collector through a window made of a material like glass or plastic. The window is designed to take advantage of the observation that sunlight is electromagnetic radiation with a distribution of frequencies. The window in a solar heat collector is transparent to incident solar radiation and opaque to infrared radiation.

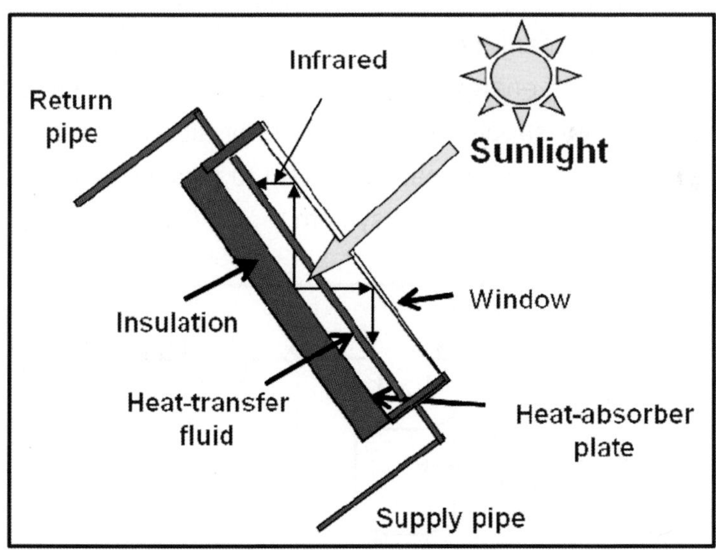

Figure 6-6. Solar Heat Collector

The heat absorber plate in the solar heat collector is a dark surface, such as a blackened copper surface, that can be heated by the absorption of solar energy. The surface of the heat absorber plate emits infrared radiation as it heats up. Sunlight enters through the window, is absorbed by the heat absorber plate, and is reradiated in the form of infrared radiation. Greenhouses work on the same principle: the walls of a greenhouse allow sunlight to enter and then trap reradiated infrared radiation. The window of the solar heat collector is not transparent to infrared radiation so the infrared radiation is trapped in the collector.

The solar heat collector must have a means of transferring collected energy to useful energy. A heat transfer fluid such as water is circulated through the solar heat collector in Figure 6-6 and carries heat away from the solar heat collector for use elsewhere.

The solar heating system sketched in Figure 6-7 uses solar energy to heat a liquid coolant such as water or anti-freeze. The heat exchanger uses heat from the liquid coolant in the primary circulation system to heat water in the secondary circulation system. The control valve in the lower right of the figure allows water to be added to the secondary circulation system. An auxiliary heater in the upper right of the figure is included in the system to supplement the supply of heat from the solar collector. It is a reminder that solar energy collection is not a continuous process. A supplemental energy supply or a solar energy storage system must be included in the design of the heating system to assure continuous availability of heat from the solar heating system.

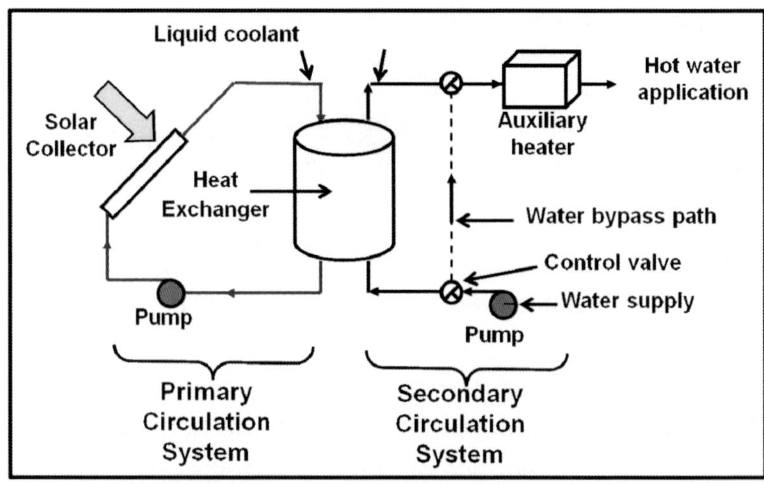

Figure 6-7. Solar Heating System [after Cassedy and Grossman, 1998, page 282]

The energy conversion efficiency of a solar heat collector is the ratio of the useful energy output by the collector to the solar energy input to the collector. The energy conversion efficiency depends on the increase in temperature relative to ambient temperature, the intensity of solar radiation, and the quality of thermal insulation. The temperature of a solar heat collector does not increase indefinitely because the window and walls of the solar heat collector cannot prevent energy from escaping by convection and radiation. The loss of energy by convection and radiation causes a decrease in energy conversion efficiency.

6.5 Solar Power Plants

Society is beginning to experiment with solar power plants and a few are in commercial operation. Solar power plants are designed to provide electrical power on the same scale as plants that rely on nuclear or fossil fuel. They use reflective materials like mirrors to concentrate solar energy. The solar power tower, the Solar Electric Generating Station in Southern California, and concentrating solar power (CSP) systems are examples of solar power plants. They are described below.

6.5.1 Solar Power Tower

Figure 6-8 is a sketch of a solar power tower with a heliostat field. The heliostat field is a field of large, sun-tracking mirrors called heliostats arranged in rings around a central receiver tower. The heliostats concentrate sunlight on a receiver at the top of the tower. The solar energy heats a fluid inside the receiver.

Figure 6-9 is a sketch of the geometry of the sun-tracking mirrors relative to the central receiving station. The heliostats must be able to rotate to optimize the collection of light at the central receiving station. Computers control heliostat orientation. As a ring of heliostats gets farther away from the tower, the separation between the ring and adjacent, concentric rings must increase to avoid shading one ring of mirrors by an adjacent ring.

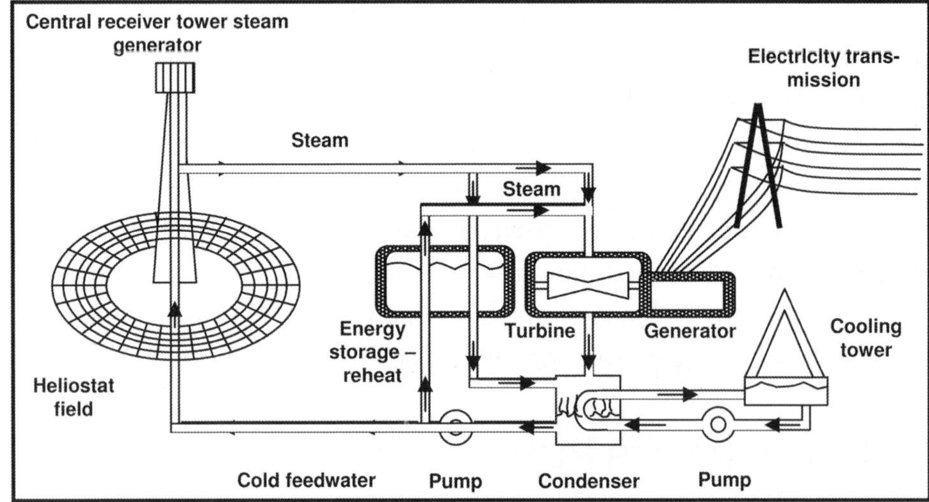

Figure 6-8. Solar Power Tower Schematic [after Kraushaar and Ristinen, 1993, page 172; and Solar Energy Research Institute (now National Renewable Energy Laboratory), Golden, Colorado]

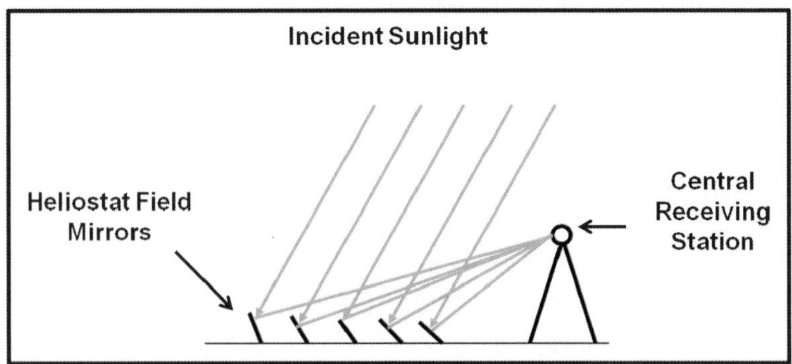

Figure 6-9. Solar Tower Sketch

The first solar power plant based on the solar power tower concept was built in the Mojave Desert near Barstow, California in the 1980's. The solar-thermal power plant at Barstow used 1900 heliostats to reflect sunlight to the receiver at the top of a 300-foot tall tower. The sunlight generates heat to create steam. The steam is used to drive a turbine or it can be stored for later use. The first solar power tower, Solar One, demonstrated the feasibility of collecting solar energy and converting it to electrical energy. Solar One was a 10 megawatt power plant. The heat transfer fluid in Solar One was steam. The Solar One installation was modified to use molten nitrate salt as the heat transfer fluid. The modified installation, called Solar Two, was able to improve heat transfer efficiency and thermal storage for the 10 megawatt demonstration project. The hot salt could be retrieved when needed to boil water into steam to drive a generator turbine.

6.5.2 Solar Electric Generating Systems

A Solar Electric Generating System (SEGS) consists of a large field of solar heat collectors and a conventional power plant. The SEGS plant in Southern California use rows of parabolic trough solar heat collectors. The collectors are sun-tracking reflector panels, or mirrors (Figure 6-10). The sunlight reflected by the panels is concentrated on tubes carrying heat transfer fluid. The fluid is heated and pumped through a series of heat exchangers to produce superheated steam. The steam turns a turbine in a generator to produce electricity.

For extended periods of poor weather, solar power plants must use auxiliary fuels in place of sunlight. A prototype SEGS plant used natural gas as an auxiliary fuel. B.Y. Goswami, et al. [2000, Section 8.7] reported that, on average, 75% of the energy used by the plant was provided by sunlight, and the remaining 25% was provided by natural gas. They further reported that solar collection efficiencies ranged from 40% to 50%, electrical conversion efficiency was on the order of 40%, and the overall efficiency for solar to electrical conversion was approximately 15%.

The overall efficiency of a solar electric generating system is the product of optical efficiency, thermal conversion efficiency, and thermodynamic efficiency. Optical efficiency is a measure of how much sunlight is reflected into the system. Thermal conversion efficiency is a measure of how much sunlight entering the system is converted to heat in the system. The thermodynamic efficiency is a measure of how much heat in the system is converted to the generation of electricity.

Figure 6-10. Solar Mirrors near Barstow, California (Fanchi, 2002)

SEGS plants are designed to supply electrical power to local utilities during peak demand periods. In Southern California, a peak demand period would be hot summer afternoons when the demand for air conditioning is high. This is a good match for a SEGS plant because solar intensity is high. Peak demand periods also correspond to periods of high pollution. One benefit of a SEGS plant is its ability to provide electrical power without emitting fossil fuel pollutants such as nitrous oxide (a component of smog) and carbon dioxide (a greenhouse gas). Other types of solar thermal plants could also be used to meet this demand. An example is the concentrating solar power concept discussed next.

6.5.3 Concentrating Solar Power

Concentrating solar power (CSP) plants use dish concentrators, solar power towers or trough concentrators to convert solar thermal energy into usable electrical energy. Solar power towers are described in Section 6.5.1 and trough concentrators, like those used in the SEGS plant, are described in Section 6.5.2. We focus on dish concentrator systems in this section. Diagrams of a dish concentrator and a trough concentrator are shown in Figure 6-11.

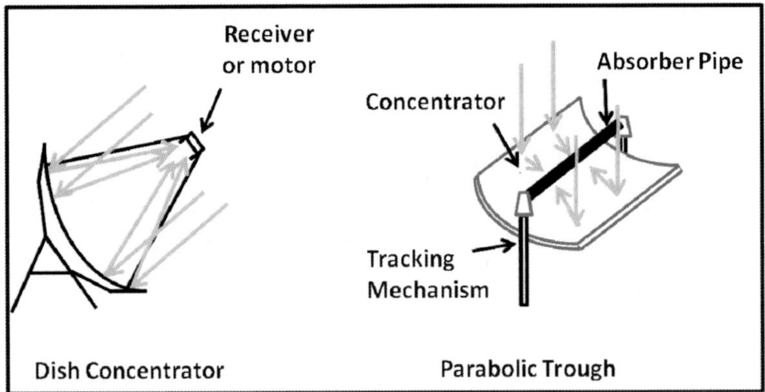

Figure 6-11. Concentrating Solar Power Systems [after Pitz-Paal, 2008]

Dish concentrator power plants are fields of individual dishes, like a satellite dish people might mount on their house for a television signal, only larger. Each dish is designed to reflect incident solar thermal energy onto a receiver. This allows each dish to function as if it were an individual power tower, with the dish itself acting as a heliostat. The benefit of this form of concentrator is that it is self-contained. Thus, a small number of dishes can be constructed in an area to provide localized power. In some cases, a single dish could even be used to provide supplemental power to a home or building. This is not feasible with a solar power tower since the size of the heliostat field must be able to provide enough power to justify the expense of constructing the project. Similarly, trough concentrators require too much infrastructure and maintenance to be built on a small scale.

A significant benefit of the dish concentrator is that each dish may move independently to track the sun. As with heliostat fields associated with solar power towers, the dish would be programmed to move on its own to follow the sun and maximize output during daylight hours. Some countries, such as Germany, view dish concentrators as being superior because they do not need to be integrated into the electrical grid, and instead can provide small scale power to isolated areas.

Point to Ponder: Why aren't solar power plants more popular in sunny, deserted areas?
One of the major advantages of solar power plants is that the energy is free. There are, however, several issues to consider that impact cost and social acceptability. The free energy from the sun has to be collected and transformed into commercially useful energy. Solar power plants to date cover the area of several football fields and produce approximately 1% of the power associated with a fossil fuel fired power plant. This means that solar power plants will cover relatively large areas and may be considered eyesores by some people. In addition, the technology of maintaining the collectors, and collecting, transforming, and transmitting solar energy is still relatively expensive. [Fanchi, 2004, Exercises 7-11 and 7-12]

6.5.4 Case Study: Spanish Solar Power

The southern coast of Spain is a sun-drenched tourist attraction. Spain has recognized that this natural resource can also be used to provide energy for itself and the members of the European Union. Consequently, Spain has placed a premium on solar power.

According to Solar PACES, an international organization that coordinates solar power projects under the auspices of the International Energy Agency, Spain first created a solar power premium payment in 2002. This premium is designed to encourage utilities to adopt solar power by granting a government subsidy for solar-generated electricity. The premium has been increased from its original level and has led to large investment in solar technology.

PS10 is a solar power tower plant near Seville, Spain, that has a capacity of 11 MW and is considered the first commercially viable solar power tower plant. Adjacent to PS10 is PS20, a solar power tower plant built after PS10 was completed. PS20 had a capacity of 20 MW and was the largest solar power tower plant in the world when it was completed. PS10 and PS20 are both projects of Abengoa Solar, a Spanish company that operates many of the solar power plants in Spain [Abengoa Solar, 2009].

Abengoa is constructing a series of connected solar power plants in the Seville area in a project called the Solúcar Platform. Along with the PS10 and PS20 plants, three solar trough plants with 50 MW capacity each have already been completed with more solar trough plants planned. In addition, an 80 MW plant is planned in the form of a concentrating solar power (CSP) plant using dish technology. The entire project is expected to be completed by 2013. An image of the area from orbit is shown in Figure 6-12 [Abengoa Solar, 2009].

Figure 6-12. NASA Image of Seville, Spain, including PS10 and PS20 [NASA Earth Observatory, 2009]

Spain is also developing a solar power tower plant called Solar Trés that will seek to use molten nitrate salt as the heat transfer fluid following the successful implementation of Solar Two in the United States. Located near Aldalusia, Spain, Solar Trés has a capacity of 17 MW and is designed to operate around the clock during the summer using molten salts for an energy storage system [Solar Spaces, 2009].

6.6 SOLAR ELECTRIC TECHNOLOGY

Solar electric technologies are designed to convert light from the sun directly into electrical energy. Some of the most important solar electric processes are the photoelectric effect and photovoltaics. These processes are introduced here.

6.6.1 Photoelectric Effect

An electrical current is the movement of electrons. Heinrich Hertz discovered in 1887 that electrons could be ejected from a metal exposed to electromagnetic radiation, but the effect depended on the frequency of the radiation. The effect was called the photoelectric effect. Albert Einstein used the concept of a quantum of energy to explain the photoelectric effect in 1905, the same year he published his special theory of relativity.

The view of light propagation at the beginning of the 20th century was that light propagates as electromagnetic waves. Einstein used the then new quantum theory to postulate that light also propagates as a particle. The particle of light is now called the photon, and is a packet, or quantum, of energy. The energy of the photon is proportional to the frequency of light. Einstein realized that a collision between a photon and an electron could transfer enough energy from the photon to the electron to eject the electron from the metal. The energy transfer required that the photon have enough energy (a high enough frequency) to overcome the work function of the metal, where the work function is the smallest energy (smallest frequency) needed to extract an electron from the metal. The work function depends on the type of material and the condition of its surface. An electron ejected from a metal because of a collision with a photon is called a photoelectron, and the photoelectric effect is the mechanism of ejecting an electron with a suitably energetic photon.

As a historical note, Einstein received the Nobel Prize for his 1905 work on the photoelectric effect, not for his work on relativity.

6.6.2 Photovoltaics

Photovoltaics is an application of the photoelectric effect. We can describe photovoltaics as the use of light to generate electrical current. We can make a photovoltaic cell, or photocell, by placing two semiconductors in contact with each other. Photocells are not sources of energy and they do not store energy. Photocells transform sunlight into electrical energy. When sunlight is removed, the photocell will stop producing electricity. If a photocell is used to produce electricity, an additional system is also needed to provide energy when light is not available. The additional system can be an energy storage system that is charged by sunlight transformed into electrical energy and then stored, or it can be a supplemental energy supply provided by another source.

Point to Ponder: What are some current uses of photovoltaics?
There are many applications of photovoltaics. They include solar powered calculators, solar powered camping equipment, solar powered satellites, the International Space Station, and solar powered cars. A solar powered traffic system for helping enforce speed limit laws is shown in Figure 6-13.

Figure 6-13. Solar Powered Traffic Control, Golden, Colorado (Fanchi, 2002)

6.7 Activities

True-False

Specify if each of the following statements is True (T) or False (F).

1. A solar heat collector is a passive solar energy device that captures sunlight and transforms radiant energy into heat energy.
2. Most of the energy that is radiated by the sun reaches the surface of the earth.
3. Passive solar technology integrates building design with environmental factors to enable the capture or exclusion of solar energy.
4. The efficiency of solar heat collectors depends on weather conditions.
5. Solar energy is inexhaustible.
6. A supplemental energy supply is often needed to accompany solar heating systems.
7. A heliostat is a large, sun-tracking mirror.
8. Biological conversion of solar energy relies on photosynthesis.
9. Photovoltaics is the use of light to generate electrical current.
10. Renewable energy is energy that is obtained from sources at a rate that is less than or equal to the rate at which the source is replenished.

Questions

1. What is a heliostat?
2. What is the difference between active and passive solar energy?
3. Which of the following systems use active solar energy?

    Roof overhang
    Solar heat collector
    Shade tree
    Solar tower
    Photovoltaic cell

4. What is the photoelectric effect?
5. How does a solar electric generating station work?
6. What causes sunlight to be reflected into space by the earth's atmosphere?

7. If a power plant consumes 250 J of energy and generates 150 J of energy, what is the energy conversion efficiency of that plant?
8. What are the benefits of dish concentrators as a form of Concentrating Solar Power?
9. What kind of power plants are the PS10 and PS20 in Spain?
10. Match the terms to their meanings by writing the letter of the meaning in the column to the left of the term.

	Term		**Meaning**
	Convection	A	Heat transfer by temperature difference between substances in contact
	Conduction	B	Transforms heat into work
	Radiation	C	Heat transfer by movement of heated substance

CHAPTER 7

RENEWABLE ENERGY – WIND ENERGY

The kinetic energy of wind and flowing water are indirect forms of solar energy and are considered renewable. Wind energy technology relies on gradients in physical properties such as atmospheric pressure to generate electrical power. Wind turbines harness wind energy and convert the mechanical energy of a rotating blade into electrical energy in a generator. The objective of this chapter is to discuss the use of wind as a source of energy for generating useful power. We begin by reviewing the history of wind power.

7.1 HISTORY OF WIND POWER

Wind has been used as an energy source for thousands of years. Historical applications include sails for ship propulsion and windmills for grinding grain and pumping water. Wind is still used today as a source of power for sailing vessels and parasailing.

The earliest known applications of wind as an energy source come from Persia [Manwell, et al., 2002]. Around 900 A.D., wind was used to drive early vertical axis windmills. Modern wind turbines are classified as either horizontal axis turbines or vertical axis turbines. A vertical axis turbine has blades that rotate around a vertical axis and its visual appearance has been likened to an eggbeater. A horizontal axis turbine has blades that rotate around a horizontal axis (see Figure 7-1). Horizontal axis turbines are the most common turbines in use today. Early vertical axis windmills had a simple design and were particularly susceptible to damage in high winds.

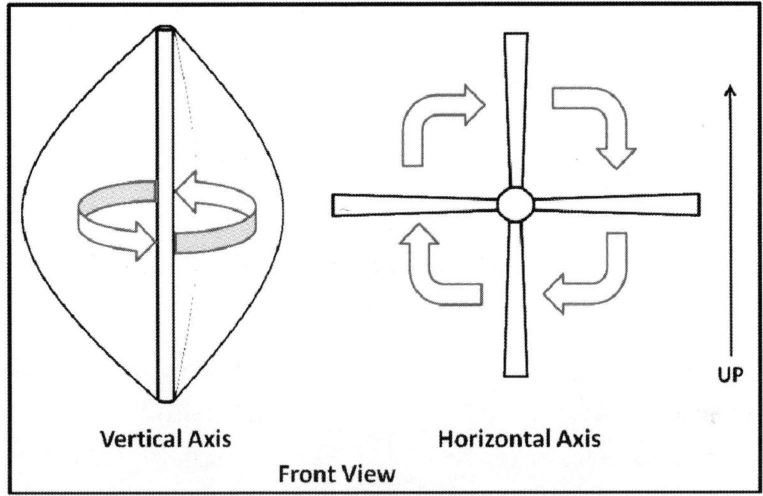

Figure 7-1. Comparison of Vertical Axis and Horizontal Axis Windmills

The first known use of horizontal axis windmills, a precursor to the technology seen predominantly today, appeared in Europe in the Middle Ages. Horizontal axis windmills could be used for a variety of simple tasks such as pumping water, grinding grain, and sawing wood. Early European windmills typically had four blades, as shown by the Dutch windmill in Figure 7-2.

Wind power was a significant contributor to European energy for centuries prior to the Industrial Revolution. As time passed, use of wind as an energy source lost favor because it was difficult to distribute and it was not always available when it was needed. Eventually coal entered the energy mix because coal had several advantages: it could be used when and where it was most needed, and helped reduce dependence on wood as a combustible fuel, which was deforesting some areas. To a lesser extent, water also overtook wind as a source of power because water could be moved through waterways such as canals and stored for more timely use in containers such as ponds.

Figure 7-2. Dutch Windmill [Fanchi, 2003]

Blades on early European windmills were usually designed to rotate together in the wind. This design made the mill susceptible to damage because of the large number of moving parts. European windmill technology eventually matured to the point that only a small part of the mill, the rotor, turned to move the blades and adjust to wind direction without affecting the adjoining structure. Controls also were implemented to allow the blades to rotate on their own and reduce the amount of operational supervision. Scientific testing of windmills led to a more sophisticated knowledge of the forces at work in a wind energy system. Many of the advances made in Europe have been integrated into modern wind turbine designs.

As use of windmills declined in Europe, a new form of windmill came into widespread use in the Western United States (Figure 7-3). This type of multi-bladed windmill, or 'fan mill', was used primarily for pumping water in the more arid regions of the United States. The water was needed for irrigation, supplying livestock with water, and providing water for steam-driven locomotives. A simple but effective regulating system was developed for these windmills that allowed them to function for long periods unattended, and set the stage for automatic control systems that are integral parts of modern turbines.

Figure 7-3. Windmill in the Western United States [Fanchi, 2009]

The first wind turbines for electricity generation appeared in the late 19th century in the United States. Following the development of the electric generator, it was only logical that someone would try to turn generator shafts using wind power. Early wind turbines were built to provide electricity for residential areas on a very small scale, usually one turbine for one home. The first larger-scale wind turbine was built by Marcellus Jacobs in the early 1920s. These turbines had three rotor blades with airfoil shapes and resemble wind turbines in common use today (Figure 7-4). The Jacobs turbine could be integrated into an electrical grid to provide energy for distribution to many consumers. A decision by the United States Rural Electrification Administration to expand the central electric grid in the 1930s may have delayed the adoption of an expanded role for wind energy in the United States energy mix. With that decision, small scale wind power lost much of its appeal since fewer areas were isolated from the electrical grid.

Figure 7-4. A Modern Wind Turbine Blade [Fanchi, 2009, Lamar, Colorado]

A large number of small-scale wind turbines with electrical production ranging from 20 kW to roughly 60 kW were developed in Denmark. Then, shortly after World War II, Johannes Juul erected the 200 kW Gedser turbine in southeastern Denmark. This turbine had several major advances built into it, including an aerodynamic design that enabled greater control of power output by varying the angle of the blade in response to changes in wind speed, and a generator that could be connected directly to the electrical grid. Also around this time, in the 1950s, German Ulrich Hütter made advances in the application of aerodynamic principles to wind turbine design. Many of Hütter's ideas led to concepts in use today. More discussion is provided by Kühn [2008] and the American Wind Energy Association [AWEA, 2009].

As stated above, the use of wind energy for electrical energy began in the late 19th century when windmills began being used on a very small scale, such as for individual houses or farms. Over the course of the 20th century, wind energy received increasing attention as a potential power source, but the technology simply did not exist for large scale use. By the middle of the century, wind energy was still limited to very small scale use despite attempts to develop the technology. Even in 1980, the total energy from wind turbines for electricity generation was less than 1000 MW, which had a small impact on the energy mix. Over the next 20 years, wind technology advanced significantly. Modern turbine technology appeared between 2000 and 2005. Today, a single wind turbine can produce up to 6 MW electrical power, and researchers are seeking to increase power output up to 10 MW using new technology, such as superconducting generators [Matthews, 2009].

7.2 Wind Turbine

A schematic of a typical modern wind turbine is shown in Figure 7-5. Moving air rotates blades attached to a generator shaft in the machine cabin. The machine cabin is known as the nacelle. It contains the electrical generator which converts the rotational energy of the rotating blades to electrical energy. Electricity is transmitted through a line in the post that connects each wind turbine to the electric grid. Therefore, the generator produces electricity that is routed directly to the electric grid.

A typical horizontal axis turbine consists of a rotor with three blades attached to a machine cabin set atop a post that is mounted on a foundation block. The machine cabin contains a generator attached to the wind turbine. The rotor blades can rotate in the vertical plane and the machine cabin can rotate in the horizontal plane.

Most modern horizontal axis turbines have three rotor blades instead of two. Rotor blades attached to a generator shaft, or rotor, make up a rigid body with a particular moment of inertia. Rotational properties of a rigid body, such as angular momentum and torque, depend on moment of inertia. The moment of inertia of the wind turbine depends on the number of rotor blades and their orientation. Turbines with two rotor blades have a higher moment of inertia when the blades are vertical than when they are horizontal. The difference in moment of inertia between the horizontal and vertical configuration of two blades introduces a mechanical imbalance that can increase wear on the system. By contrast, the use of three equally spaced blades adds the cost of another blade but allows a symmetric placement of blades that makes it easier to balance the blades as they rotate. The improved stability of turbines with three rotor blades increases wind turbine reliability and reduces maintenance costs.

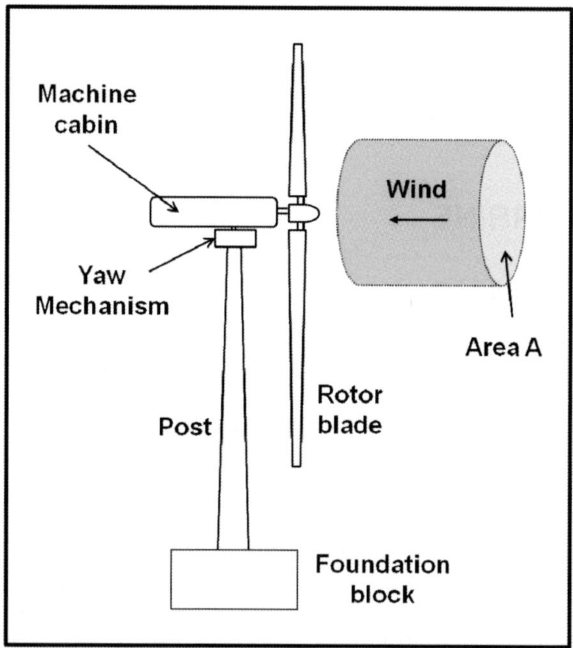

Figure 7-5. Schematic of a Wind Turbine

Figure 7-5 shows a yaw mechanism attached to the post just below the machine cabin. The purpose of the yaw orientation system is to align the rotor shaft with wind direction. Figure 7-6 illustrates the difference between yaw, pitch and roll in aeronautical terms. Yaw refers to rotation of the nose in the horizontal plane around the vertical axis. Rotation of the airplane around the B to B' line is called pitch, and rotation of the wings around the A to A' line is called roll.

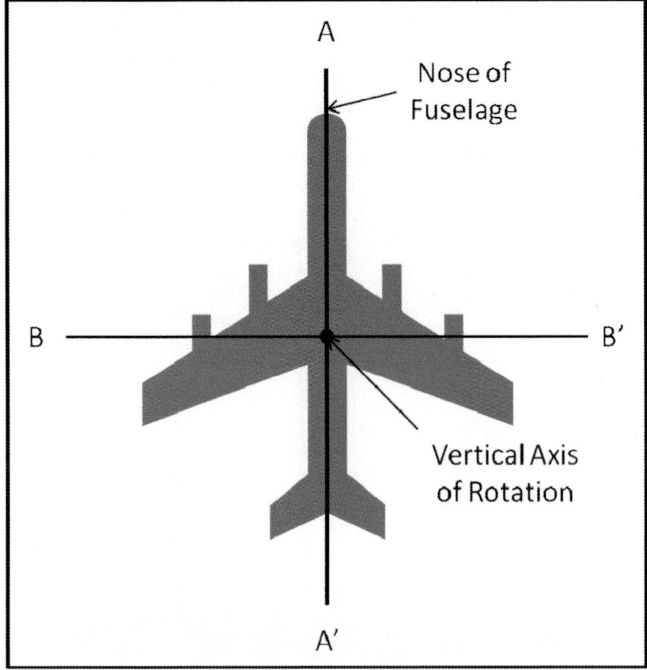

Figure 7-6. Aeronautical Orientation System

Modern wind turbines are several hundred feet tall. For example, a wind turbine that generates 1.6 megawatts of electrical power is approximately 113 meters (370 feet) tall from its base to the tip of the rotor blade. The storage tanks in Figure 7-7 illustrate the scale of a modern wind turbine and demonstrate that wind turbines can be erected on existing industrial properties. For example, wind turbines have been erected near fossil fuel fired power plants, along highways, on ranchland, and in shallow waters offshore.

If the speed of rotation of the tip of the rotor blade is fast enough, it can be lethal to birds entering the fan area of the rotor blade. This environmental hazard can be minimized by selecting locations for wind turbines that avoid migration patterns. Another way to minimize environmental impact is to use large rotor blades that turn at relatively low speeds of rotation.

Figure 7-7. Modern Wind Turbine near Rotterdam, Holland (Fanchi, 2003)

7.2.1 Turbine Power Output

Wind power can be maximized when wind direction is perpendicular to the plane of rotation of the rotor blades. A change in wind direction can put stress on wind turbines. In addition, wind speed is seldom constant; it can vary from still to tornado or hurricane speed.

Typical wind turbines have three regions of power output, illustrated in Figure 7-8, that depend on wind speed. Region I occurs when the wind speed is low and little or no power is produced. Region II occurs with intermediate wind speed allowing the turbine to begin producing meaningful power output. Region III occurs when the wind speed is high and the turbine reaches maximum power. The power output at Region III will peak at a predetermined level known as the capacity of the wind turbine. Eventually, if the wind speed is great enough, the turbine will reach a cutout point at which time it shuts down entirely. This is generally reserved for instances such as hurricanes or tornadoes, where the wind speed is dangerous to the equipment.

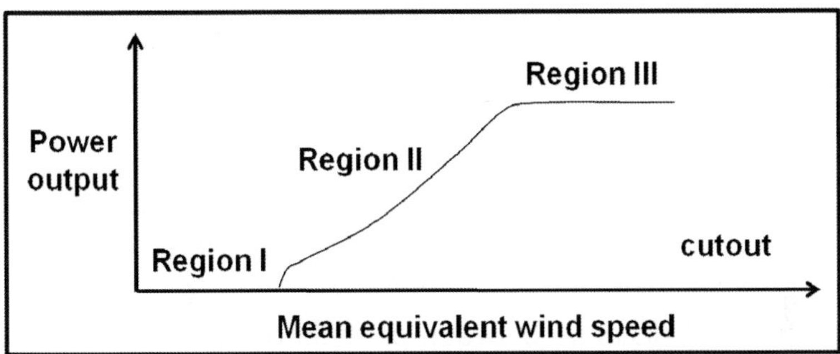

Figure 7-8. Wind Turbine Power Output Regions

One factor that affects the efficiency of a wind turbine is the efficiency of converting mechanical energy of the rotor blade into electrical energy. In 1928, Albert Betz showed that the maximum percentage of wind power that can be extracted is approximately 59.3% of the power in the wind. Another factor affecting electrical power output from a wind turbine is the reliability of the wind turbine. The rate of rotation of the rotor blade depends on wind speed and size of the blades. If the wind speed is too great, the rotor blade can turn too fast and damage the system.

7.3 Wind Farms

A wind farm or wind park is a collection of wind turbines. The areal extent of the wind farm depends on the radius R of the rotor blades and the effective radius R_{eff} of the wind turbine (Figure 7-9). A wind turbine must have enough space around the post to allow the fan of the rotor blade to face in any direction. The minimum spacing between the posts of two equivalent wind turbines must be $2R_{eff}$ to avoid collisions between rotor blades. If we consider the aerodynamics of wind flow, which is the factor that controls turbine spacing, the turbine spacing in a wind farm increases to at least 5 to 10 times rotor diameter 2 R [Sørensen, 2000, page 435] behind the plane of the rotor blade.

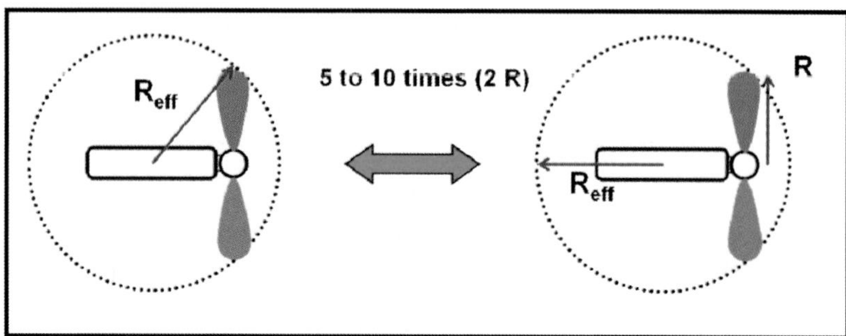

Figure 7-9. Wind Turbine Spacing

The additional distance between posts is designed to minimize wind turbulence between wind turbines and enable the restoration of the wind stream to its original undisturbed state after it passes by one turbine on its way to the next turbine. Wind turbine spacing, or the distance between turbines, is an important factor in determining the surface area, or footprint, needed by a wind farm. Figure 7-10 shows the spacing of wind turbines in a Texas wind farm near Sweetwater, Texas. The train on the left side of the figure shows the scale of the wind turbines.

Figure 7-10. Wind Turbine Spacing in a Texas Wind Farm [Fanchi, 2009]

7.3.1 Offshore Wind Farms

Current technology favors onshore wind farms as the most economical form of wind farm. There are drawbacks to onshore wind farms that can be mitigated by construction of offshore wind farms [US EIA website, 2009, Annex XXIII]. For example, the sight of a 200-300 foot tall wind turbine does not appeal to everyone, and wind turbines do make some noise. Offshore wind turbines can be placed beyond visible and audible range of many populated areas.

Several countries around the world do not have good conditions for onshore wind farms but do have good conditions along nearby shoreline. Some conditions that would hinder the development of onshore wind farms include unreliable onshore wind patterns or a population density so large that access to land is too costly. Wind currents tend to be stronger and smoother offshore than on land. Onshore topographic features like mountains, valleys, and skylines can disrupt the smooth flow of air and introduce turbulence. Offshore wind farms can be built to provide power to densely populated coastal areas where the cost to access land is prohibitive.

European countries with coastlines began constructing offshore wind turbines in the early 1990s, led by Denmark and the United Kingdom. The wind resource map in Figure 7-11 shows that the Gulf Coast, Atlantic coast, Pacific coast, and Great Lakes of the United States are areas where wind conditions are better offshore than onshore. Offshore wind conditions give these areas access to wind power.

A concern that is particularly significant to the United States is the cost of transporting large, heavy components to isolated parts of the country. Transportation costs for development of offshore wind farms can be lower because marine shipping and handling equipment is better suited for the heavy components needed for wind farms.

Offshore wind farms can have many benefits compared to onshore wind farms, but there are a number of drawbacks. One issue is the difficulty of creating an infrastructure system to get the electricity from the turbine to shore without significant loss of power. This concern is mitigated when onshore wind farms are located a significant distance from the end user. In that case, the cost of transmitting power over long distances is comparable for both onshore and offshore power generation by wind farms.

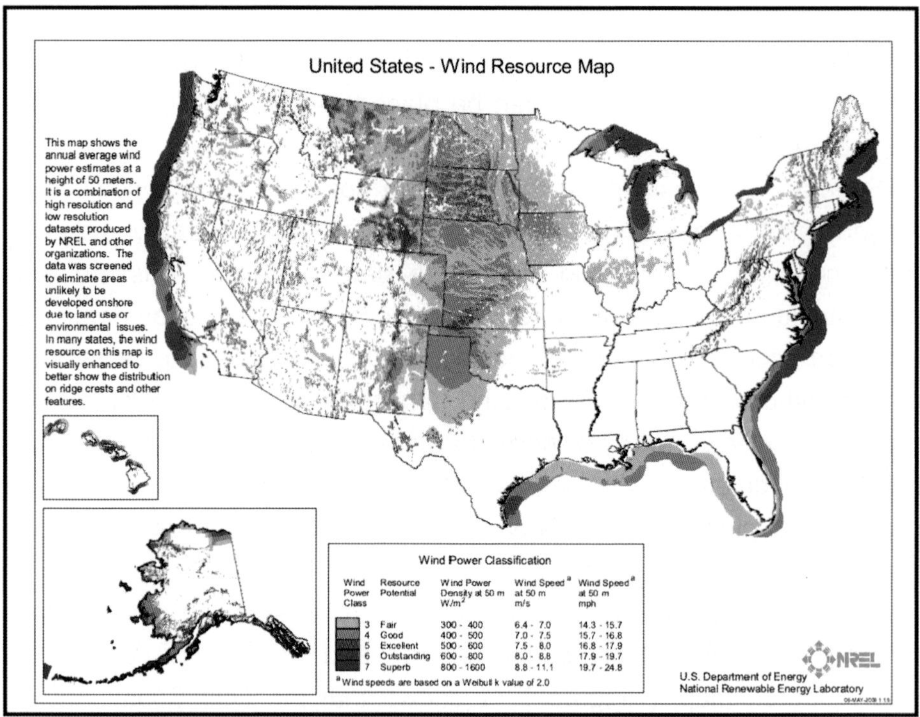

Figure 7-11. Wind Patterns in the United States (NREL, accessed 11-5-2009)

Another drawback of offshore wind farms is increased maintenance costs. Costs are greater for offshore facilities than facilities on land for several reasons. Salt in seawater has a corrosive effect that increases water damage to equipment. The location of turbines offshore increases the difficulty of accessing them. Offshore turbines are subjected to more severe weather conditions, including strong wave activity, which increases the risk of serious damage to equipment.

Countries with sea coasts adjacent to shallow sea shelves include Denmark, the Netherlands and the United Kingdom. Shallow water can extend more than 12 miles from shore. These countries have a geographic advantage for placement of offshore wind farms because the cost of building an offshore wind farm in shallow water is less than the cost of building an offshore wind farm in deep water. Two major difficulties of constructing an offshore wind farm in deep water are installation and control of wind turbines that are adequately anchored to the sea floor. Section 7.3.2 examines an option for limiting this difficulty.

Point to Ponder: Would you want a wind farm in your backyard?

The first offshore wind farm to be built in the United States is proposed for the Nantucket Sound area of Massachusetts. The wind farm was approved by the U.S. Department of the Interior in April 2010. The farm, called Cape Wind by farm sponsors, will consist of 130 wind turbines constructed over five miles from the nearest shoreline in a region called Horseshoe Shoal. According to CapeWind.org, the Shoal "has strong, consistent winds; is located in protected shallow water; has close proximity to landfall and electrical interconnections; and is out of way of shipping lanes and commercial boating traffic." The farm is designed to produce 420 MW of energy. Each turbine would be spaced 1800 to 2700 feet apart, which is enough space for easy navigation of the surrounding waters by fishing craft and shallow water boats.

So what is the problem? The largest concern exhibited by the local population is that the aesthetic impact will drastically affect the value of beachfront homes. A related concern is that the drop in home values would lead to a drop in tax revenue gathered from property taxes. CapeWind.org answered these concerns by pointing to a similar wind farm constructed at Tunoe Knob, Denmark in 1995. The concerns of Tunoe Knob residents were similar to those held by Massachusetts residents. Tunoe Knob home values remain steady fifteen years after construction of the wind farm, and the presence of the wind farm is no longer a major topic of discussion.

The Cape Wind project is expected to create jobs in the area and have a positive impact on the local economy. Opponents of the project argue that it could seriously damage the natural beauty of Nantucket Sound, a popular tourist destination, which in turn could reduce tourism, tourism-related jobs, and harm the local economy.

Projects such as Cape Wind can provide clean, renewable energy, create energy-related jobs, and increase energy independence. They can also have social and environmental consequences that need to be considered.

7.3.2 Advances in Offshore Wind Technology

An example of an advanced offshore wind turbine is shown in Figure 7-12. Some offshore wind turbines are designed to float and maneuver in accordance with wind and wave patterns, while others are designed to be moored to the sea floor. Floating offshore wind turbines can be used in

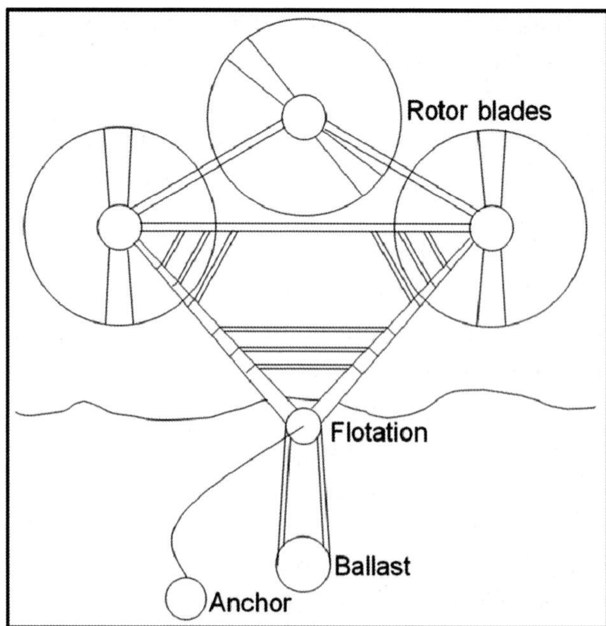

Figure 7-12. Floating Offshore Wind Turbine

waters that are too deep for conventional turbines. Another benefit of the turbine shown in Figure 7-12 is the use of multiple sets of rotor blades. A floating turbine can move to align with wind direction, so multiple sets of blades can be mounted without risk of the blades interfering with each other. The wind turbine must be sturdy enough to withstand bad weather conditions, and there must be a way to transmit power to the electrical grid. Power transmission costs and transmission line power losses are too great for commercial installation of advanced offshore wind turbines.

7.4 Case Study: European Wind Power

Over the past decade, the feasibility and economics of wind have caused a groundswell of support for wind as an alternative to fossil fuels. Technological advances in wind turbine technology, electrical transmission technology and power grid organization have improved the efficiency and cost effectiveness of wind dramatically. With these improvements, nations lacking fossil fuel resources have begun viewing wind as a possible primary energy source for the future. While the United States is taking large strides in this direction, Europe has become a leader in wind power. The following discussion is based on a study by the European Wind Energy Association [EWEA, 2009].

Many members of the European Union have a high UN Human Development Index score. We showed in Chapter 1 that quality of life is correlated to energy use. Few European countries have access to domestic fossil fuels, and those that do fear depletion of these resources as well as the environmental impact that use of fossil fuels may cause. In an attempt to secure a local energy resource with a fixed fuel cost, these countries have turned to wind energy. Wind is an abundant natural resource that is renewable and virtually free. Over time, the only cost of operating a wind power plant is the cost of operations and maintenance because the fuel, wind, is free. Since 1995, the European Union has maintained greater wind power capacity than the rest of the world combined [EWEA, 2009, Figure 3.1].

Denmark was among the first nations in the world to begin producing large amounts of wind power and to include wind as a major contributor to their national power supply. By the end of 2008, 20.3% of Denmark's power supply was provided by wind power, which is the largest percentage of any European country. Spain (12.3%) and Portugal (11.4%) are the only other countries over ten percent [EWEA, 2009, Figure 3.10]. Overall, the 27 member countries of the European Union use wind to meet 4.1% of total electricity demand.

The lack of available open space in Europe has led to many advances in offshore wind power technology. By land area, Denmark is one of the smallest countries in the European Union, yet Denmark is a leader in wind power because of its offshore wind capacity. Until 2008, when overtaken by the United Kingdom, Denmark had long been the world leader in offshore wind capacity. Denmark had 409 MW of its 3,180 total MW of wind power from offshore wind [EWEA, 2009, Table 3.2]. At the end of 2008, there were nine offshore wind farms in the world, all nine in Europe.

Table 7.1
Total Installed Offshore Wind Capacity

Country	MW of Offshore Wind Power (End 2008)
United Kingdom	591
Denmark	409
Netherlands	247
Sweden	133
Belgium	30
Ireland	25
Finland	24
Germany	12
Italy*	0
*At the end of 2008, Italy had one 0.08 MW test turbine offshore but it was not connected to the electric grid.	

Of the 1,471 MW of offshore wind capacity installed in the EU, all but 35 MW of that capacity has been installed since the beginning of 2001. This demonstrates the exponential growth in wind power, and particularly offshore wind power, in this region [EWEA, 2009, Figure 3.9]. Similarly, over that same period of time, total EU wind power capacity grew from 12,887 MW to 64,935 MW. While it should be noted that the EU has grown from 15 to 27 members in that time frame, 63,857 MW, or 98%, of the total EU wind capacity comes from the 15 members that have been present since 2000 or earlier.

The European Union is one of many international organizations taking steps to formulate a plan for future power production and use. The 2009 EU Renewable Energy Directive was created as a means to force EU Member States to begin actively addressing their energy future. The Directive required each EU member to estimate its energy consumption for both renewable and non-renewable energy sources each year from 2010 to 2020. This consumption must include the three factors that use the most energy: heating/cooling, electricity, and transportation. The Directive requires every country to produce at least 12% of their total power from wind by 2020, although the relatively low cost of wind in today's energy market may enable countries to exceed this number.

Many of the wind power projections for EU members are ambitious. One example is France, which produced 3,404 MW of power from wind at the end of 2008. France is expected to increase its wind capacity to at least 23,000 MW by 2020. This capacity requires adding 1,633 MW each year. The 23,000 MW total includes 4,000 MW of new capacity in offshore wind. France must receive assistance from its neighbors to design the technology for offshore wind farms and create the infrastructure necessary to deliver this power to consumers.

Another concern is availability of wind turbines, cables, and other equipment needed to install significantly more wind capacity. Today, manufacturing capacity is able to meet demand and provide equipment at a reasonable price. The European plan calls for a significant increase in wind power capacity in a short period of time, and this increase could affect the price and availability of equipment and trained personnel.

7.5 CAN WIND PROVIDE ALL OF OUR ENERGY NEEDS?

Wind appears to have many of the advantages that would make it an appealing solution to our energy problems. When properly designed and located, wind turbines are environmentally benign. Some people may object to large fields filled with wind turbines, but fields of wind turbines can be built on property to serve a dual purpose. For example, wind farms can be built on West Texas ranch land or along the roadways in Rotterdam, Holland. In both cases the turbines rotate well above activities below.

Point to Ponder: How many wind turbines would we need to supply global energy demand?

We can estimate the number of wind turbines that would be needed to supply global energy demand by making a few key assumptions. For the sake of argument, let us assume that the world population in year 2100 will be eight billion people and the amount of energy needed to provide each person an acceptable quality of life will be 200,000 megajoules per year. Both of these assumptions can be challenged, but they define a specific scenario so that we can estimate the number of wind turbines needed for the scenario.

Suppose we use wind turbines that can provide 4 mw each. We would need about 12.7 million wind turbines to supply global energy demand in 2100. If these wind turbines are collected in wind farms that can provide 1000 megawatts per wind farm, we would need approximately 50,700 wind farms [Fanchi, 2004, Exercises 15-7 and 12-8]. If we assume the turbine radius is 108 feet and assume the area occupied by each wind farm is approximately square, with a turbine separation of about ten times turbine radius, we estimate that each wind farm will occupy about nine square miles. All of the farms would occupy an area of about 465,000 square miles, or about 16% of the area in the contiguous United States. The area is smaller than the state of Alaska, but larger than the state of Texas.

We can build enough wind farms to provide the energy we need based on area and power capacity, as illustrated in the Point to Ponder above. The ability to build wind farms offshore further reduces concerns about access to area. Still, other issues must be considered. As discussed in Section 6.2, wind like solar power, is intermittent and is not always available. We need to provide energy when the wind does not blow. We need to distribute the energy where it is needed and when it is needed. We need to provide energy in a form that best fits the need. We need to be willing to accept the environmental impact of wind farms, including their appearance and impact on wildlife.

Some unresolved issues are technical, others are social. An example of a technical issue is the question of how to provide energy on demand, even when the wind is not blowing. There are several options to consider. For example, we could use wind energy to charge batteries, develop large scale energy storage capacity, or use wind energy to produce hydrogen for use in fuel cells. One advantage of hydrogen production is that we could use hydrogen in the transportation sector. Hydrogen is considered in detail in Chapter 10.

Much work remains to be done to identify the optimum strategy for providing renewable energy. It is important to note that wind power in the United States cost approximately US$0.04 to US$0.06 per kilowatt hour in 2009, which was comparable to the cost of electricity from natural gas. The cost-competitiveness of wind and the demand for cleaner energy by consumers is encouraging the growth of wind energy around the world.

Figure 7-13 shows the exponential growth of wind production in the United States. This figure can be misleading because it gives the impression that wind is a major contributor to the United States energy mix. To

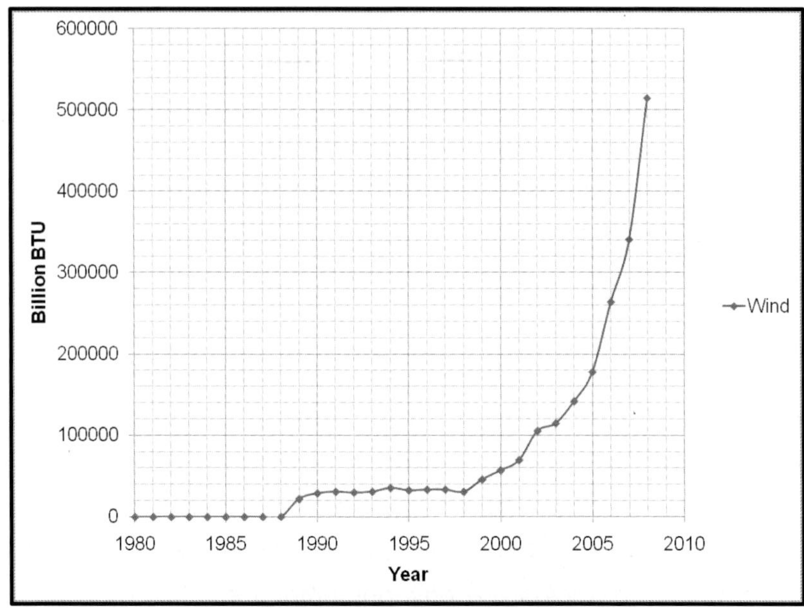

Figure 7-13. Energy Production from Wind in the United States

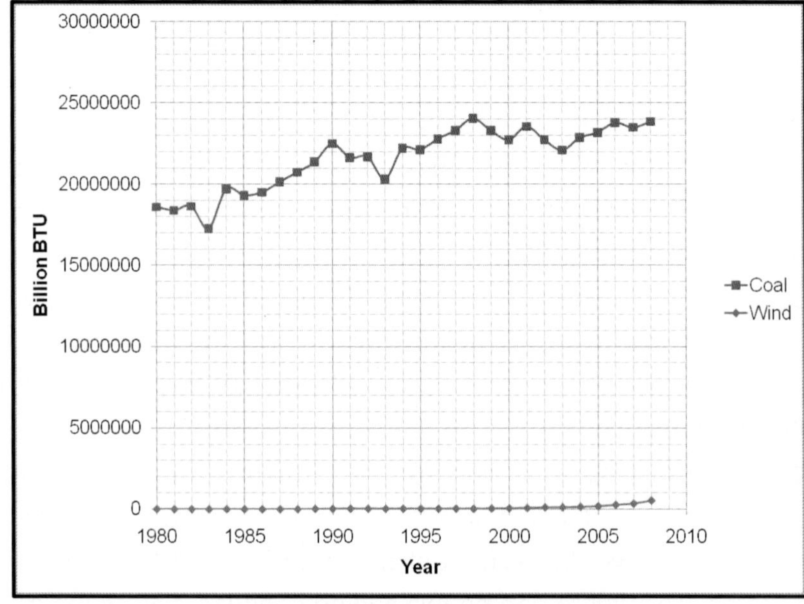

Figure 7-14. Comparison of Energy Production from Wind and Coal in the United States

understand the actual role of wind in the United States, we compare the production of energy from wind and coal in Figure 7-14. The relative contribution of energy production from wind is much less than the contribution of energy production from coal. Wind energy contributes a larger percentage of energy in some European countries than in the United States, but wind energy is still in its infancy.

7.6 ENVIRONMENTAL IMPACT

Wind energy is a renewable energy that is considered a clean energy because it has a minimal impact on the environment compared to other forms of energy. Wind turbines provide electrical energy without emitting greenhouse gases. On the other hand, we have already observed that the harvesting of wind energy by wind turbines can have environmental consequences.

Rotating wind turbine blades can kill birds and interfere with migration patterns. Wind turbines with slowly rotating, large diameter blades and judicious placement of wind turbines away from migration patterns can reduce the risk to birds. Wind farms can have a significant visual impact that may be distasteful to some people. Wind turbines produce some noise when they operate. In the past, wind turbines with metal blades could interfere with television and radio signals. Today, turbine blades are made out of composite materials that do not interfere with electromagnetic transmissions.

An aesthetic concern that is coming to the forefront as more wind turbines are placed in populated areas is a concept known as "flicker." Flicker is the effect a wind turbine has on the area of land in its shadow. Turbine blades that rotate in daylight cast a shadow that causes a flickering light effect. This flicker can be irritating or distracting to people if their homes or businesses are in the shadow of the turbine.

7.7 Activities

True-False

Specify if each of the following statements is True (T) or False (F).

1. Modern wind turbines provide energy whenever the wind is blowing.
2. Solar energy and wind energy are intermittent sources of energy.
3. A sailboat can use wind energy and water currents to move.
4. Wind turbines must be spaced to minimize turbulence.
5. Two drawbacks of wind energy are the visual impact and footprint of wind farms.
6. A horizontal axis turbine has blades that rotate around a horizontal axis.
7. A wind farm is a collection of wind turbines.
8. Wind turbines have no adverse effect on the environment.
9. Use of wind as an energy source is a recent development (within the last 200 years).
10. The United States was the first country to construct an offshore wind farm.

Questions

1. A renewable energy project will generate 400 MW of wind power using 100 wind towers and will cost approximately US$ 600 million.
    a. On average, how much wind power will be provided by each wind tower?
    b. What is the average cost per MW?
2. What is the minimum distance between two wind turbine posts on a wind farm?
3. Does a wind turbine have any negative environmental consequences?
4. When and where were horizontal axis windmills first utilized?
5. Where were fan mills developed and what were their purpose?
6. What does the yaw mechanism on a wind turbine do?
7. Why do most modern wind turbines use three blades?
8. List three benefits of offshore wind farms relative to onshore wind farms.

9. Germany had 23,903 MW of wind capacity at the end of 2008. If Germany must reach 49,000 MW by the end of 2020, how many MW must be installed each year?
10. What is “flicker”?

CHAPTER 8

RENEWABLE ENERGY – ENERGY FROM WATER

Water is a renewable energy source that may have a large impact on the future energy mix, and it is already contributing to worldwide energy today. About 34% of the renewable energy consumed by the United States in 2008 was hydroelectric (Appendix B), while almost 80% of the renewable energy consumed by the world was hydroelectric (Appendix C). Hydropower depends on the movement of water in the water cycle illustrated in Figure 8-1.

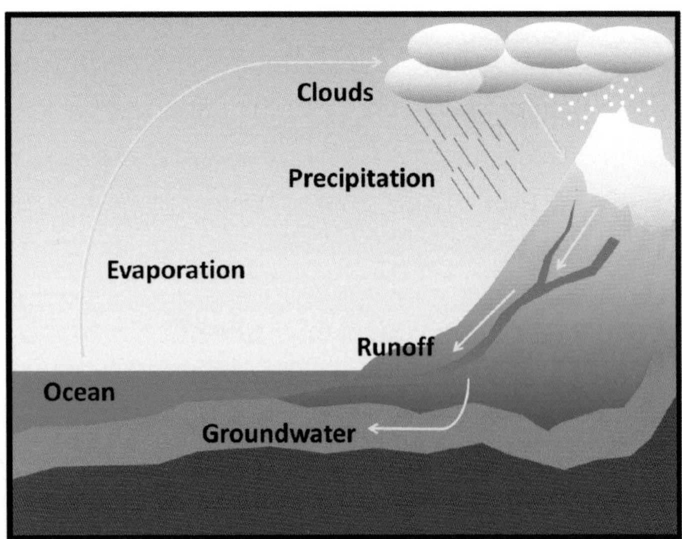

Figure 8-1. The Water Cycle

The water cycle is a global cycle of moving water. Water evaporates from lakes and oceans and rises into the atmosphere where it coalesces into clouds. Clouds can move over all parts of the earth until atmospheric pressure and temperature changes lead to water precipitation in the form of rain or snow. Some of the precipitated water seeps into the earth as groundwater, and some flows along rivers and streams back to lakes and oceans, where the water cycle begins again. The hydrosphere includes groundwater and water found in oceans, glaciers, surface waters such as rivers and lakes, and atmospheric moisture.

Hydropower harvests energy from the water cycle and is considered renewable since the water cycle is driven by energy from the sun in a seemingly endless series of cycles. Hydropower is usually obtained from streams and rivers. Water in other systems can be used to generate power. For example, energy can be harvested from waves and tides, and from temperature differences in columns of water. We consider power production using energy from water in this chapter.

8.1 Hydroelectric Power

People have known for some time that falling water could be used to generate electric power. We summarize the history of hydroelectric power and describe the generation of hydroelectric power here.

8.1.1 History of Hydroelectric Power in the United States

Hydropower was used by the Greeks to turn water wheels for grinding wheat into flour more than 2,000 years ago. French hydraulic and military engineer Bernard Forest de Bélidor wrote *Architecture Hydraulique*, a four-volume work published between 1730 and 1770. Bélidor presented hydraulic principles and described vertical- and horizontal-axis machines. Hydroelectric power provided approximately 40% of the electricity generated in the United States in 1940. Today hydroelectric power provides approximately 7% of the electricity generated in the United States. Table 8-1 highlights key events in the history of hydroelectric power in the United States.

Table 8-1
History of Hydroelectric Power in the United States
[Source: DOE Hydropower History, 2009]

Year	Comment
1775	U.S. Army Corps of Engineers founded.
1880	Grand Rapids Electric Light and Power Company in Michigan provided electricity from a dynamo belted to a water turbine to power lamps for theatre and storefront lighting.
1881	Niagara Falls city street lamps powered by hydropower.
1882	World's first hydroelectric power plant began operation on the Fox River in Appleton, Wisconsin.
1886	About 45 water-powered electric plants in U.S. and Canada.
1887	San Bernardino, California opens first hydroelectric plant in the west.
1889	Two hundred electric plants in the U.S. use waterpower for some or all electricity generation.
1901	Federal Water Power Act enacted.
1902	Bureau of Reclamation established.
1920	Federal Power Commission established.
1933	Tennessee Valley Authority established.
1935	Federal Power Commission covers hydroelectric utilities engaged in interstate commerce.
1937	The first Federal dam, the Bonneville Dam, begins operation on the Columbia River. Bonneville Power Administration established.

The United States Corps of Engineers was founded in 1775 with the naming of a Chief Engineer in the Continental Army. Its mission has been to provide engineering services for military and public projects that are in the national interest, which often includes canals, dams, and levees. The United States Federal Water Power Act, now known as the Federal Power Act, was enacted in 1901 to coordinate hydroelectric projects. The Bureau of Reclamation was established in 1902. The Bureau was formerly known as the Reclamation Service and oversees water resource management in the United States. It is part of the United States Department of Interior. The Federal Power Commission was established by the Federal Power Act in 1920 with the authority to issue licenses for hydro development on public lands. The Tennessee Valley Authority was established in 1933 during the Great Depression. The authority of the Federal Power Commission was extended in 1935 to include hydroelectric projects built by utilities engaged in interstate commerce.

8.1.2 Generation of Hydroelectric Power

Many of the first commercial electric power plants relied on flowing water as their primary energy source. A schematic of a hydroelectric power plant is presented in Figure 8-2. Water flows from an upper elevation to a lower elevation through a pipeline called a penstock. The change in elevation of the water is a change in the potential energy of the water. Dams with turbines and generators convert the change in potential energy into mechanical kinetic energy. The water current turns a turbine that is connected to a generator. The turbine is called the prime mover because it rotates the generator shaft. Mechanical energy of falling water is trans-

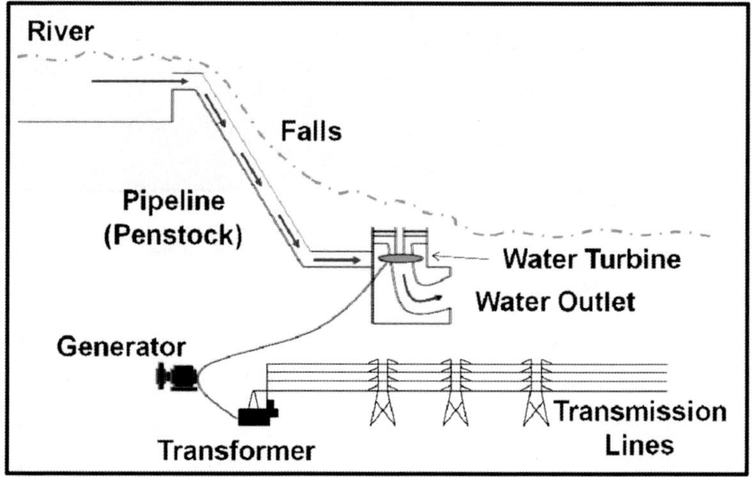

Figure 8-2. Principles of Hydroelectric Power Generation

formed into the kinetic energy of rotation of the turbine. The rotating turbine rotates a shaft that is connected to either a magnet adjacent to a coil of wire or a coil of wire adjacent to a magnet. The relative motion of the wire and the magnet produces an alternating current. The current generator converts mechanical energy into electrical energy.

A dam is built by diverting or blocking water flow past the planned location of the dam. When the dam is completed, the water is impounded, or blocked, so that water begins to pool and form a reservoir upstream of the dam. The elevation, or head, is the height the water falls from upstream of the dam to downstream. The head is replaced by the effective head for realistic systems. The effective head is less than the actual head, or elevation, because water flowing through a conduit such as a pipe will lose energy to friction and turbulence. The rate that water falls through the effective head depends on the volume of the penstock shown in Figure 8-2. If the penstock volume is too small, the output power will be less than optimum because the flow rate could have been larger. On the other hand, the penstock volume cannot be arbitrarily large because the flow rate through the penstock depends on the rate that water fills the reservoir behind the dam.

Figure 8-3. Reservoir behind Hoover Dam (Fanchi, 2002)

Figure 8-3 shows the reservoir behind the Hoover Dam at the Arizona-Nevada border in the United States. The light colored rock in the figure bounding the reservoir shows that the depth of the reservoir is below normal. The drainage region into the reservoir had experienced an extended drought at the time the photo was taken.

The volume of water in the reservoir and corresponding height depends on the water flow rate into the reservoir. During drought conditions, the elevation can decline because there is less water in the reservoir. During rainy seasons, the elevation can increase as more water drains into the streams and rivers that fill the reservoir behind the dam. Hydroelectric power facilities must be designed to balance the flow of water through the electric power generator with the water that fills the reservoir through such natural sources as rainfall, snowfall, and drainage. Three hydroelectric power plant sizes are shown in Table 8-2. The plant sizes are classified by electrical power generating capacity. Hoover dam on the border of Nevada and Arizona has an installed capacity of over 1000 MW, which makes it a large hydroelectric power plant.

Table 8-2
Hydroelectric Power Plant Sizes
[DoE Hydropower, 2002]

Size	**Electrical Power Generating Capacity (MW)**
Micro	< 0.1
Small	0.1 to 30
Large	> 30

Figure 8-4 shows the recent decline in hydroelectric power production in the United States. The contribution of hydroelectric power to the United States energy mix has begun to decline as public awareness of the environmental impact of dams has increased.

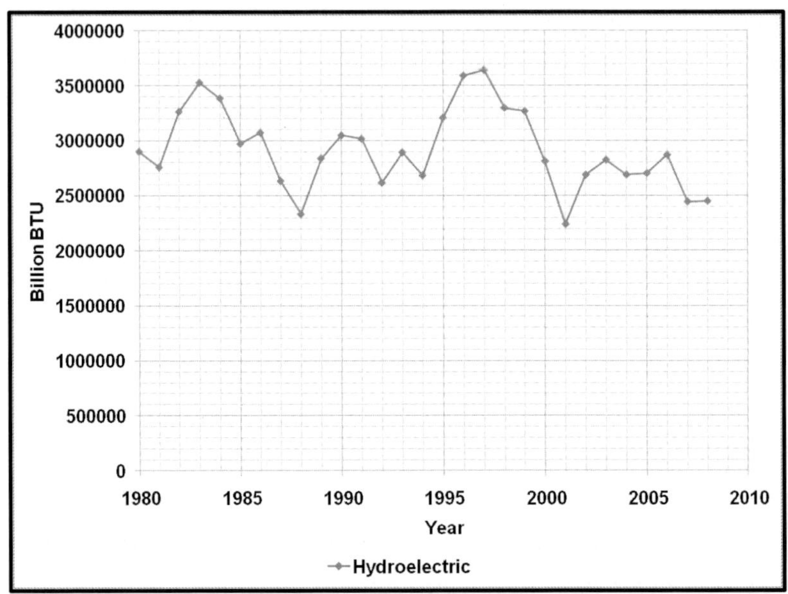

Figure 8-4. Hydroelectric Power Production in the United States

Adverse environmental effects include flooding upstream of the dam, and changing river flow downstream of the dam. These effects occur wherever a dam is built. The Case Study below illustrates some adverse environmental effects of hydroelectric power for a recent hydroelectric power project in China.

8.2 Case Study: Three Gorges Dam

The Three Gorges Dam is the world's largest hydroelectric dam built to date [International Rivers, 2009; CTGP, 2009]. Built on the Yangtze River in China, dam construction began in 1994 and was completed in late 2009. At completion, the Three Gorges Dam is 101 meters (330 feet) high and 2,309 meters (7,575 feet) wide. Its reservoir extends 660 kilometers (410 miles) behind the dam. The power plant portion of the dam went on-line in October 2008 with 26 generators. The plant provides a total capacity of 18,200 MW, roughly equivalent to ten to fourteen modern nuclear power plants or 22 large coal-fired plants. Six additional generators are planned. Official figures state the final cost of the dam at approximately US$27.2 billion, although unofficial cost estimates are as high as US$88 billion.

Supporters of the Three Gorges Dam project consider it to be a symbol of China's technological and economic progress. The dam replaces the burning of 30 million tons of coal every year, which helps decrease China's carbon output and creates a renewable source of energy for China's expanding economy. The dam reduces the potential for downstream floods on the Yangtze, a problem that has historically plagued inhabitants along the river's shore. Control over the river's flow makes river navigation easier.

Since its inception, the Three Gorges Dam has been highly controversial. The Yangtze River is among the world's longest rivers, and the Yangtze River valley has been heavily populated for thousands of years. When a river of this magnitude is dammed, the water must be stored somewhere. The reservoir formed behind the Three Gorges Dam forced relocation of the populations of 13 cities, 140 towns and 1,350 villages. Artifacts and ruins from thousands of years of Chinese history were submerged by the reservoir.

The Chinese government promised to provide land for relocated farmers and jobs for relocated urban workers, but arable land is difficult to come by in heavily populated China. Many of the jobs that were going to be provided by state-run enterprises did not appear. Some cities built to house resettled populations have grown and thrived, but others have become destitute with resettled populations falling into poverty. It has also been alleged that funds allocated for resettling populations have been embezzled by local officials.

The expected environmental impact of the Three Gorges Dam is already being seen, but some unexpected issues are emerging. The dam has changed the flow of the river, with upstream water becoming stagnant while the flow speed has increased in some downstream waterways. The river used to carry large loads of silt that would mitigate its speed, but that silt is being collected at the dam and the downstream river speed has increased causing damage to levees. The benefit of controlling unpredictable flooding has been undermined by this damage.

The change in water speed and composition is also causing changes in the chemical levels and acidity (pH) of the river. These changes are harming fish populations. Migratory fish are affected by the dam, and many indigenous fish species have become endangered. Trash and debris are collecting upstream of the reservoir. Downstream, seawater is beginning to force its way up the river because the river's flow is not as strong as it was previously. This seawater is threatening farm lands and drinking water supplies near the mouth of the river.

The dam and associated reservoir may be impacting seismic activity in the area. The change in the distribution of weight on the surface of the earth changes subsurface stresses. Larger dam systems seem to have the greatest negative effect. The Three Gorges Dam is constructed to be resistant to forces of earthquakes in the area, but office buildings, homes and schools in the dam's vicinity are not.

The reservoir created by the dam is less stable than anticipated. Much of the land surrounding the Yangtze River is made of silt and dirt deposits from millions of years of river floods and recessions. This land is not as stable as the rock canyons surrounding dams such as the Hoover Dam in the United States. Five months after the Three Gorges Dam was impounded, 150 geological events were recorded according to the Chinese business magazine *Caijing*. These events included landslides caused by erosion. There is concern that the number of events will increase and cause damage to the population around the reservoir basin. Additional relocation may be necessary.

Point to Ponder: What is the environmental impact of a hydroelectric dam?

While there are many environmental hazards associated with hydroelectric dams, many of which have only recently been observed and studied, there are still many benefits. With greenhouse gas emissions a primary environmental concern in today's society, hydroelectric power is a viable alternative to fossil fuels. According to Wengenmayr [2008a, page 25], "Reservoirs can – depending on their geological and climatic situation – emit carbon dioxide and, as a result of the decomposition of plant material, also methane. But even when the production of the materials for constructing a typical hydroelectric plant is taken into account, on the average only the equivalent of a few grams of CO_2 per kilowatt hour are released, [Stefan] Hirschberg [from the Paul Scherrer Institut in Villigen, Switzerland] explains. An average coal-fired power plant, in contrast, emits a kilogram (approx. 2 lb) of CO_2 per kilowatt of power produced, and a Chinese power plant emits even as much as 1.5 kg (approx. 3 lb) per KW."

Wengenmayr went on to say that "[t]he outdated Chinese power plants also lack filtering systems. In densely-populated areas, they shorten the lifespan of the population measurably. According to Hirschberg's studies, 25,000 years of life expectancy are lost there per gigawatt of power produced per year. Furthermore, the coal mines degrade the overall ecological and social conditions in China. They emit enormous amounts of methane, and only between 1994 and 1999, more than 11,000 miners lost their lives in accidents."

The Three Gorges Dam project has become an example for developing countries seeking to increase their energy independence by building local hydroelectric power plants. Developing nations without the economic flexibility or natural resources of the United States view hydroelectric power as a virtually carbon-free, renewable and domestic energy option and China is viewed as the new world leader in this industry.

The Chinese people appear to be divided on the Three Gorges project. Current government officials seem eager to distance themselves from the project as evidence of the environmental and geological impact of the dam continues to surface. On the other hand, the dam is a technological marvel in its size and scope and it reflects the technological and economic prosperity of the Chinese nation. It also provides an enormous source of renewable, virtually carbon-free energy that decreases the country's reliance on coal and other environmentally unfriendly energy sources.

8.3 Waves and Tides

Ocean waters are solar-powered sources of energy. The mechanical energy associated with waves and tides, and thermal energy associated with temperature gradients in the ocean can be used to drive electric generators. We consider each of these energy sources in this section.

8.3.1 Waves and Tides

In our discussion of hydropower, we saw that water can generate power when it moves from a high potential energy state to a low potential energy state, like a waterfall. If the moving water is an ocean wave, the elevation varies sinusoidally with time. The sinusoidal wave has crests and troughs. The wave amplitude, or height of the wave relative to a calm sea, depends on weather conditions: it will be small during calm weather and can be very large during inclement weather such as hurricanes. The build-up of an ocean wave off the coast of Oahu is shown in Figure 8-5. The change in the potential energy of wave motion can be transformed into energy for performing useful work.

Figure 8-5. Motion of an Ocean Wave, Oahu, Hawaii (Fanchi, 2003)

The energy density of waves breaking on a coastline in favorable locations can average 65 megawatts/mile (40 megawatts/kilometer) of coastline [DoE Ocean, 2002]. The motion of the wave can be converted to mechanical energy and used to drive an electricity generator. The wave energy can be captured by floats or pitching devices like the paddles presented in Figure 8-6. If we lay a paddle on the water, the paddle should be buoyant enough to move up and down with the wave. The change in potential energy of the paddle depends on the amplitude of the water wave, the size of the paddle, and the density of the paddle. Paddles can be anchored to the seabed to constrain their range of movement.

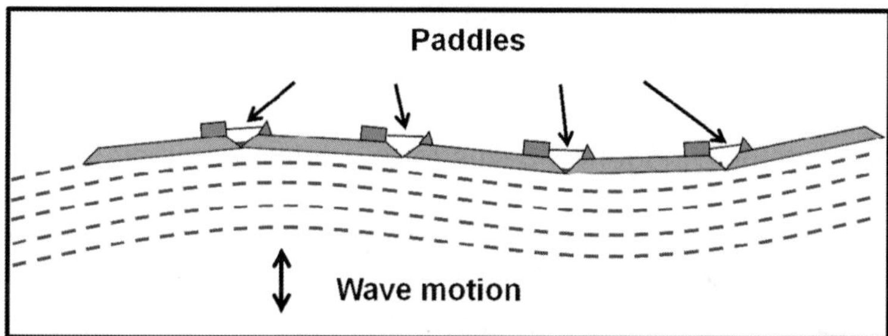

Figure 8-6. Capturing Wave Energy [after Shepherd and Shepherd, 1998, page 227; and McVeigh, 1984]

Another approach to capturing wave energy is to install an oscillating water column (OWC). The rise and fall of water in a cylindrical shaft drives air in and out of the top of the shaft. The motion of the air provides power to an air-driven turbine. The output power depends on the efficiency of converting wave motion to mechanical energy and then converting mechanical energy to electrical energy. An OWC system is presented in Figure 8-7.

A third approach to capturing wave energy is the wave surge or focusing technique. A tapered channel installed on a shoreline concentrates the water waves and channels them into an elevated reservoir. Water flow out of the reservoir is combined with hydropower technology to generate electricity.

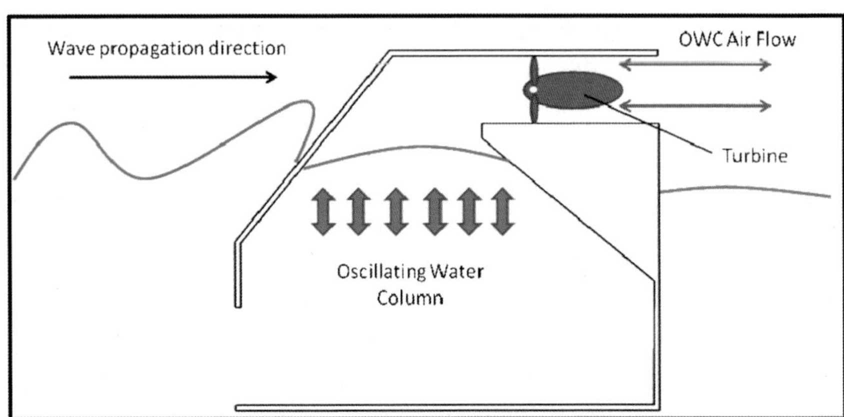

Figure 8-7. Oscillating Water Column (OWC) Wave Energy System [after Graw, 2008, page 78]

The ebb and flow of tides produces tidal energy that can be captured to produce electricity. Figure 8-8 illustrates a tidal energy station.

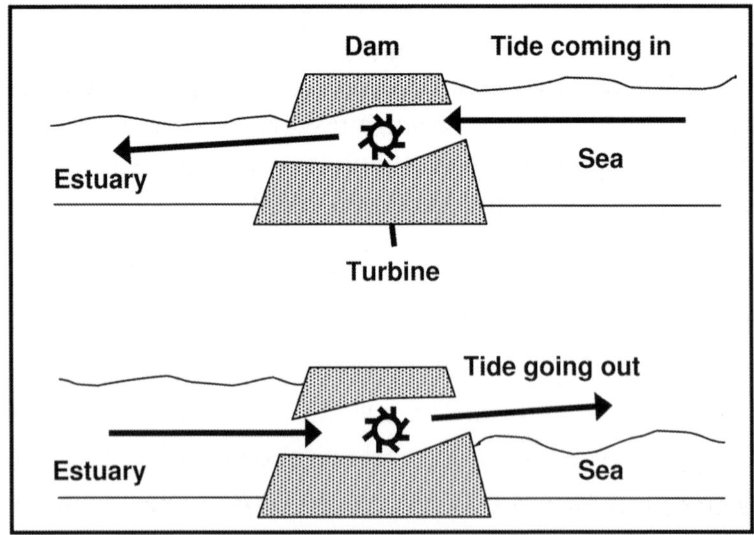

Figure 8-8. Capturing Tidal Energy [after Shepherd and Shepherd, 1998, page 227; and *Renewable Energy – A Resource for Key Stages 3 and 4 of the UK National Curriculum*, Renewable Energy Enquires Bureau, Oxfordshire, United Kingdom]

A dam with a sluice is erected across the opening to a tidal basin to capture tidal energy. A sluice is a channel with a gate that can be used to control the flow of water. The sluice is opened to let the tide flow into the basin and then closed when sea level drops. The water can flow through the gate four times per day because the tide rises and falls twice a day. Hydropower technology can be used to generate electricity from the elevated water in the basin.

Tides can rise as high as 15 meters on the Rance River in France. The Rance tidal energy station is capable of generating 240 megawatts of power [DoE Ocean, 2002; Serway and Faughn, 1985, page 114]. Tidal energy stations can have an environmental impact on the ecology of the tidal basin because of reduced tidal flow and silt buildup.

Point to Ponder: How much energy can be obtained from coastal waves?

A typical fossil fuel fired power plant produces 1000 megawatts of power. If we assume that the average wave power along a coast is 30 megawatts per kilometer and assume the efficiency of power production is 25%, the equivalent of 133 kilometers (82 miles) of coastline is needed to produce 1000 megawatts of power [Fanchi, 2004, Exercise12-6]. The length of coastline can be reduced by placing two or more lines of paddles perpendicular to the coast.

8.3.2 Ocean Thermal

Solar energy warms the world's lakes, seas and oceans. The temperature of sea water near the surface can be much warmer than the temperature of sea water in deep oceans and seas, especially in deep water near the equator. The difference in temperature at different depths is called a temperature gradient. Temperature gradients near ocean bottom geothermal vents and underwater volcanoes exist between hot water near the heat source and cooler waters away from the heat source. If the temperature gradients are large enough, they can be used to generate power using ocean thermal energy conversion (OTEC) power plants.

Temperature gradients in deep oceans between 20° North latitude and 20° South latitude are most suitable for OTEC systems [Plocek, et al., 2009]. This area around the globe includes the Caribbean Sea, Gulf of Mexico, Atlantic Ocean, Pacific Ocean, Indian Ocean, and Arabian Sea. Differences in temperature between warm surface water and cooler deep water can be over 20°C (40°F).

Three types of OTEC systems are recognized [DoE Ocean, 2002]: closed-cycle plants, open-cycle plants, and hybrid plants. A closed-cycle plant circulates a heat transfer fluid with a low boiling point like ammonia through a closed system. The fluid is heated with warm seawater until it is flashed to the vapor phase. The vapor is routed through a turbine and then condensed back to the liquid phase with cooler seawater.

An open-cycle plant uses pressure changes to flash warm seawater to vapor and then use the vapor to turn a turbine. An example of a pressure change that can transform liquid water to vaporized water is a vacuum chamber. Liquid water at atmospheric pressure can be placed in the vacuum chamber and flashed to the vapor phase.

A hybrid plant is a combination of an open-cycle plant with a heat transfer fluid. Warm seawater is flashed to steam in a hybrid plant and then used to vaporize the heat transfer fluid. The vaporized heat transfer fluid circulates in a closed system as in the closed-cycle plant.

An ocean thermal energy conversion system can be built on land near a coast, installed near the shore on a continental shelf, or mounted on floating structures for use offshore. In an open-cycle plant or a hybrid plant, the flashed seawater can be condensed as fresh, desalinated water while the salt and other contaminants are captured. In some parts of the world, a desirable byproduct of an OTEC system is this production of desalinated water. Salt from the desalination process must be disposed of in a manner that minimizes its environmental impact.

8.4 GEOTHERMAL

Geothermal energy is energy provided by the Earth. The Earth's interior is subdivided into a crystalline inner core, molten outer core, mantle, and crust. Basalt, a dark volcanic rock, exists in a semi-molten state at the surface of the mantle just beneath the crust. Drilling in the Earth's crust has shown that the temperature of the crust tends to increase linearly with depth. The interior of the Earth is much hotter than the crust. The source of heat energy is radioactive decay, and the crust of the Earth acts as a thermal insulator to prevent heat from escaping into space.

Heat energy harvested from geological sources is called geothermal energy. Figure 8-9 shows how thermal energy from lava creates steam on the southern coast of Hawaii. The dark surface is a solidified lava field that is cool enough to walk on but contains tubes that channel molten rock to the sea.

Figure 8-9. Lava meeting the Pacific, Hawaii (Fanchi, 2002)

Geothermal energy is obtained by using a fluid (such as hot water or steam) to carry the internal energy of the Earth to the surface of the Earth. Geothermal energy can be obtained from temperature gradients between the shallow ground and surface, subsurface hot water, hot rock several kilometers below the Earth's surface, and magma. Magma is molten rock in the mantle and crust that is heated by the large heat reservoir in the interior of the Earth. In some parts of the crust, magma is close enough to the surface of the Earth to heat rock or water in the pore spaces of rock. Magma, hot water, and steam are carriers of energy.

Some of the largest geothermal production facilities in the world are at the Geysers in California, and in Iceland. Figure 8-10 shows the Blue Lagoon and Svartsengi geothermal power plant in Keflavik Iceland. Surplus warm water from the plant fills up the Blue Lagoon, which is a bathing resort. The power plant is on the far side of the lagoon. Water vapor can be seen in the cooler air. Geothermal plants at the Geysers and Svartsengi are built adjacent to geothermal energy sources. The technology for converting geothermal energy into useful heat and electricity can be categorized as geothermal heat pumps, direct-use applications, geothermal heating systems, and geothermal power plants. Each of these technologies is discussed below.

Figure 8-10. Svartsengi Blue Lagoon, Iceland (A. Fanchi, 2006)

8.4.1 Geothermal Heat Pump

A geothermal heat pump uses energy near the surface of the Earth to heat and cool buildings. The temperature of the upper three meters

(10 feet) of the Earth's crust remains in the relatively constant range of 10° Centigrade to 16° Centigrade. A geothermal heat pump for a building consists of ductwork in the building connected through a heat exchanger to pipes buried in the shallow ground nearby. The building can be heated during the winter by pumping water through the geothermal heat pump. The water is warmed when it passes through the pipes in the ground. The resulting heat is carried to the heat exchanger where it is used to warm air in the ductwork. During the summer, the direction of heat flow is reversed. The heat exchanger uses heat from hot air in the building to warm water that carries the heat through the pipe system into the cooler shallow ground. In the winter, heat is added to the building from the Earth, and in the summer heat is removed from the building.

8.4.2 Direct-Use Applications

A direct-use application of geothermal energy uses heat from a geothermal source directly in an application. Hot water from the geothermal reservoir is used without an intermediate step such as the heat exchanger in the geothermal heat pump. Hot water from a geothermal reservoir may be piped directly into a facility and used as a heating source. A direct-use application for a city in a cold climate with access to a geothermal reservoir is to pipe the hot water from the geothermal reservoir under roads and sidewalks to melt snow.

Minerals that are present in the geothermal water will be transported with the hot water into the pipe system of the direct-use application. Some of the minerals will precipitate out of the water when the temperature of the water decreases. The precipitate will form a scale in the pipes and reduce the flow capacity of the pipes. Filtering the hot water or adding a scale retardant can reduce the effect of scale. In either case, the operating costs will increase.

8.4.3 Geothermal Heating System

An example of a geothermal application with a heat exchanger is the system sketched in Figure 8-11.

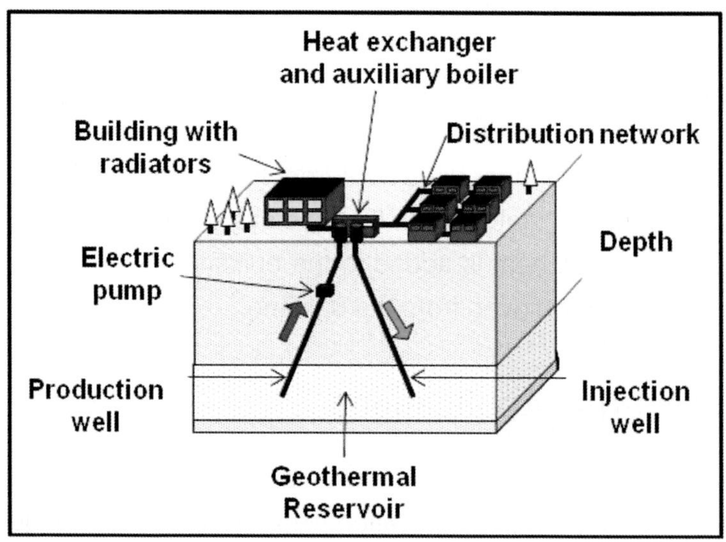

Figure 8-11. Geothermal Heating System [after Shepherd and Shepherd, 1998, page 149; and *Renewable Energy – A Resource for Key Stages 3 and 4 of the UK National Curriculum*, Renewable Energy Enquires Bureau, Oxfordshire, United Kingdom]

The geothermal reservoir is an aquifer with hot water or steam. The production well is used to withdraw hot water from the geothermal reservoir and the injection well is used to recycle the water. Recycling helps maintain reservoir pressure. If the geothermal reservoir is relatively small, the recycled, cooler water can lower the temperature of the aquifer. The electric pump in the figure is needed to help withdraw water because the reservoir pressure in this case is not high enough to push the water to the surface. Heat from the geothermal reservoir passes through a heat exchanger and is routed to a distribution network.

8.4.4 Geothermal Power Plants

Geothermal power plants use steam or hot water from geothermal reservoirs to turn turbines and generate electricity. Dry steam power plants use steam directly from a geothermal reservoir to turn turbines. Flash steam power plants allow high-pressure hot water from a geothermal reservoir to flash to steam in lower-pressure tanks. The resulting steam is used to turn turbines. A third type of geothermal power plant called a binary-cycle plant uses heat from moderately hot geothermal water to flash a second fluid to the vapor phase. The second fluid must have a lower boiling point than water so that it will be vaporized at the lower temperature associated with the moderately hot geothermal water. There must be enough heat in the geothermal water to supply the latent heat of vaporization needed by the secondary fluid to make the phase change from liquid to vapor. The vaporized secondary fluid is then used to turn turbines.

8.4.5 Managing Geothermal Reservoirs

Like oil and gas reservoirs, the hot water or steam in a geothermal reservoir can be depleted by production. The phase of the water in a geothermal reservoir depends on the pressure and temperature of the reservoir. Single-phase steam will be found in low pressure, high temperature reservoirs. In high-pressure reservoirs, the water may exist in the liquid phase, or in both the liquid and gas phases, depending on the temperature of the reservoir. When water is produced from the geothermal reservoir, both the pressure and temperature in the reservoir can decline. In this sense, geothermal energy is a non-renewable, finite resource unless the produced hot water or steam is replaced. A new supply of water can be used to replace the produced fluid or the produced fluid can be recycled after heat transfer at the surface. If the rate of heat transfer from the heat reservoir to the geothermal reservoir is slower than the rate of heat extracted from the geothermal reservoir, the temperature in the geothermal reservoir will decline during production. To optimize the performance of the geothermal reservoir, it must be understood and managed in much the same way that petroleum reservoirs are managed.

8.4.6 Hot, Dry Rock

Another source of geothermal energy is hot, dry rock several kilometers deep inside the Earth. Figure 8-12 illustrates a hot, dry rock facility that is designed to recycle an energy carrying fluid.

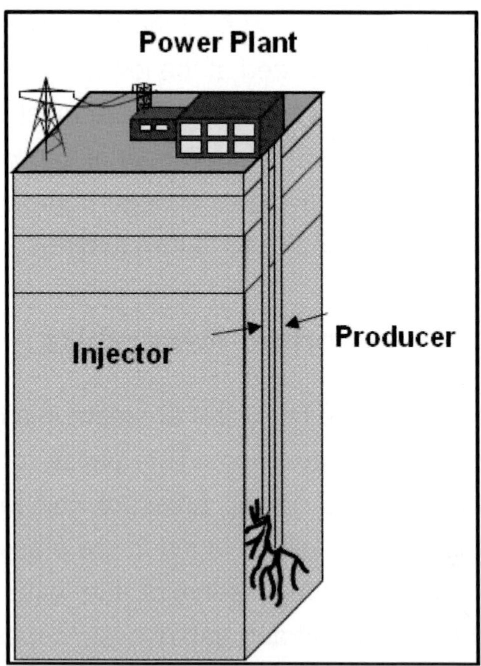

Figure 8-12. Geothermal Energy from Hot, Dry Rock

Hot, dry rocks are heated by magma directly below them and have elevated temperatures, but they do not have a means of transporting the heat to the surface. In this case, it is technically possible to inject water into the rock, let it heat up and then produce the hot water.

Water is injected into fissures in the hot, dry rock through the injector and then produced through the producer. The power plant at the surface uses the produced heat energy to drive turbines in a generator. After the hot, produced fluid transfers its heat to the power plant, the cooler fluid can be injected again into the hot, dry rock.

> **Point to Ponder: Can we use water to provide our energy needs?**
> Water is available worldwide and has been used to provide energy for centuries. It does have limitations, however. Hydropower is already widely used and can impact the environment. Geothermal energy can be a commercial source of energy in the locations where it is accessible. Waves and tides can provide commercial energy, but the efficiency of energy production is relatively low and the environmental impact, including the visual impact along coasts, can be significant.
>
> Although water appears to have a limited ability to satisfy our energy needs, it does contain an element that has great potential: hydrogen. We discuss the role that hydrogen may play in the future energy mix in Chapter 10.

8.5 ACTIVITIES

True-False

Specify if each of the following statements is True (T) or False (F).

1. OTEC technology relies on the motion of water waves.
2. Hydropower is power generated by the change in potential energy of water moving through a gravitational field.
3. An OTEC system must be mounted on floating structures.
4. Dams built between the sea and estuaries can be used to generate electricity from tidal motion.
5. OTEC systems use temperature gradients to generate power.
6. Increasing the volume of the penstock in a hydroelectric power plant will always increase the power output.
7. The Three Gorges Dam is the world's largest hydroelectric dam.
8. Construction of the Three Gorges Dam had a significant financial, environmental and societal cost.
9. Water is used to turn a turbine in an OWC system.

10. A geothermal heat pump uses energy near the surface of the Earth to heat and cool buildings.

Questions

1. What is hot, dry rock geothermal energy?
2. How is energy brought to the surface from a hot, dry rock geothermal source?
3. Can a drought affect the performance of a dam?
4. Is the energy associated with waves and tides mechanical energy or thermal energy?
5. What is the purpose of the United States Corps of Engineers?
6. What are four types of energy involved in hydroelectric power generation?
7. List three advantages and disadvantages associated with the Three Gorges Dam.
8. How do paddles capture wave energy?
9. Can geothermal reservoirs be depleted by production?
10. What part of the world is most suitable for OTEC systems?

CHAPTER 9

RENEWABLE ENERGY – BIOENERGY AND SYNFUELS

Biomass refers to wood and other plant or animal matter that can be burned directly or can be converted into fuels. Figure 1-2 in Chapter 1 shows that wood was the energy source for virtually all of the energy consumed in the United States until the middle of the 19th century. Wood was replaced by other energy sources, notably fossil fuels, hydroelectricity, and nuclear energy until today biomass is a relatively small contributor to the energy mix. Biomass provided approximately 53% of the renewable energy produced in the United States in 2008 (Appendix B). To put this in perspective, renewable energy provided approximately 10% of the energy produced in the United States in 2008. Consequently, biomass was about 5% of United States energy production in 2008.

Technologies exist to convert plants, garbage, and animal waste into natural gas. Synthetic fuels (synfuels) are fossil fuel substitutes created by chemical reactions using such basic resources as coal or biomass. Synthetic fuels are used as substitutes for conventional fossil fuels such as natural gas and oil. We discuss biomass and synfuels in this chapter.

9.1 BIOMASS

People have used biomass for fuel ever since we learned to burn wood. Biomass is matter that was recently formed as a result of photosynthesis. Photosynthesis is a process that converts solar energy into chemical energy. The process uses sunlight as an energy source to synthesize carbohydrates from carbon dioxide and water. Photosynthesis occurs in green plants, and some organisms such as bacteria and algae.

Biomass includes wood and other plant or animal matter that can be burned directly or can be converted into fuels. In addition, products derived from biological material are considered biomass. Methanol, or wood alcohol, is a volatile fuel that has been used in race cars for years. Another alcohol, clean burning ethanol, can be blended with gasoline to form a blended fuel (gasohol) and used in conventional automobile engines, or used as the sole fuel source for modified engines.

Availability is one advantage biomass has relative to other forms of renewable energy. Energy is stored in biomass in the solid, liquid, or gaseous state, and can be stored until needed. Biomass may be grown, harvested, stored, and converted into a more useful form. Other renewable energy forms, such as wind and solar energy, depend on environmental conditions that can vary considerably and are not always available when they are needed.

Figure 9-1. A Compost Heap on a Farm Near Houston, Texas (Fanchi, 2003)

Biomass energy is typically converted using chemical reactions to produce heat during combustion. Technologies exist to convert plants, garbage, animal dung, and municipal solid waste (MSW) into natural gas.

One example of a biomass to energy conversion is the decay of organic waste in a compost heap (Figure 9-1). The decay releases heat and methane. Another example of a biomass to energy conversion process is the production of gas from decay of organic waste in a landfill.

A landfill is a pit filled with garbage. When the pit is full, we can cover it with dirt and insert a pipe through the dirt into the pit. The pipe provides a conduit for producing the natural gas that is generated from the decay of biological waste in garbage. The landfill gas is filtered, compressed, and routed to the main gas line for delivery to consumers. Figure 9-2 shows a landfill gas skid with a flare pipe. The ridge on the right hand side of the picture is the top of the landfill, and the dark features on the ridge are pieces of a material that are used to cover the landfill and prevent gas from escaping into the atmosphere.

Figure 9-2. Landfill Gas Skid with Flare Near Denver, Colorado (Fanchi, 2004)

9.1.1 Energy Density

The amount of energy that can be produced from biomass depends on the heat content of the material when it is dry. Table 9-1 shows energy densities for several common materials. Energy density is the amount of energy in a material divided by the amount of material. Historically, energy density was one of the most important factors considered in selecting a fuel. Coal and oil have relatively large energy densities and were often preferentially chosen as the raw fuel that was input to power plants.

Table 9-1
Energy Density of Common Materials*

Material	**Energy Density**	
	MJ/kg	**MJ/m^3**
Crude oil	42	37,000
Coal	32	42,000
Dry wood	12.5	10,000
Hydrogen, gas	120	10
Hydrogen, liquid	120	8,700
Methanol	21	17,000
Ethanol	28	22,000
*Source: Sørensen, 2000, page 473 and 552		

The amount of material in energy density can be expressed as mass or volume. For example, a kilogram (2.2 pounds mass) of crude oil provides 42 megajoules of energy, while one kilogram of sawmill scrap provides 12.5 megajoules of energy. Thus, a kilogram of crude oil provides over three times as much energy during the combustion process as the same mass of sawmill scrap. An example that illustrates the relative value of fossil fuels to biomass is the production of heat by natural gas burners compared to a boiler that relies on the combustion of wood. The natural gas burner is more efficient than a boiler because the energy density of natural gas is greater than the energy density of wood [Sørensen, 2000, page 473]. The energy density of biomass is less than fossil fuel.

Some types of biomass used for energy include forest debris, agricultural waste, wood waste, animal manure, and the non-hazardous, organic portion of municipal solid waste. A major component of municipal solid waste is sewage. Developing countries are among the leading consumers of biomass because of their rapid economic growth and increasing demand for electricity in municipalities and rural areas.

Point to Ponder: Why is energy density important?
Energy density tells us how much energy we can obtain from a given amount of material and it is a quantity we can use to compare different fuels. For example, a larger energy density in fossil fuels than wood was one of the reasons wood was replaced by coal and oil as a fuel in the 19th century. A smaller volume of fossil fuel could provide more energy than wood, which meant less storage was needed for fossil fuel. This was an especially valuable advantage in transportation, since reducing the storage volume for fuel on ships and trains decreased the cost of transporting goods and passengers.

9.1.2 Wood

Wood has historically been a source of fuel. We have already learned that wood was a primary energy source through much of history [Nef, 1977], but deforestation became such a significant problem in 16th century England that the English sought, and found, an alternative source of fuel: coal. Today, many people rely on wood as a fuel source in underdeveloped parts of the world. Economic sources of wood fuels include wood residue from manufacturers, discarded wood products, and non-hazardous wood debris.

An increased reliance on wood as a fuel has environmental consequences, such as pollution, an increased rate of deforestation and an increase in the production of byproducts from wood burning. Figure 9-3 shows the afternoon sun as it was seen through smoke from a forest fire. One way to mitigate the environmental impact is to implement reforestation programs. Reforestation programs are designed to replenish the fo-forests. Research is underway to genetically engineer trees that grow fast, are drought resistant, and easy to harvest. Fast-growing trees are an example of an energy crop, that is, a crop that is genetically designed to become a competitively priced fuel.

Figure 9-3. Afternoon Sun Obscured by Smoke Near Denver, Colorado (Fanchi, Summer 2002)

We know from observation and the chemistry of wood combustion that burning wood consumes carbon, hydrogen and oxygen. The combustion reactions produce water, carbon monoxide, and carbon dioxide, a greenhouse gas. Greenhouse gases are gases that capture sunlight reradiated as heat from the earth. Proponents of climate change believe an increase in the amount of greenhouse gases in the atmosphere will increase the capture of reradiated sunlight and lead to harmful changes in the climate. The emission of greenhouse gases produced by the combustion of biomass has an environmental impact that needs to be considered when comparing different energy sources.

9.1.3 Ethanol

Ethanol is an alcohol that can be added to gasoline to increase the volume of fuel available for use in internal combustion engines. A biofuel is a fuel that contains biomass. Ethanol is made from a biomass feedstock. A feedstock is the raw material supplied to an industrial processor. Residues that are the organic byproducts of food, fiber, and forest production are economical biomass feedstock. Examples of residue include sawdust, rice husks, corn stover, and bagasse. Corn stover is used to produce ethanol. It is a combination of corn stalks, leaves, cobs, and husks. Bagasse is the residue that is left after juice has been extracted from sugar cane. A biomass research facility at the National Renewable Energy Laboratory in Golden, Colorado is shown in Figure 9-4. A fermentation tank is in the foreground of the figure.

Figure 9-4. Biomass Research Facility, National Renewable Energy Laboratory, Golden, Colorado (Fanchi, 2003)

The process of ethanol production proceeds in several steps. First, the feedstock must be delivered to the feed handling area. The biomass is then conveyed to a pretreatment area where it is soaked for a short period of time in a dilute sulfuric acid catalyst at a high temperature. The pretreatment process liberates certain types of sugars and other compounds. The liquid portion of the mixture is separated from the solid sludge and washed with a base such as lime to neutralize the acid and remove compounds that may be toxic to fermenting organisms. A beer containing ethanol is produced after several days of anaerobic fermentation. The beer is distilled to separate ethanol from water and residual solids.

The production of ethanol from biomass requires distillation. During distillation, a mixture is boiled and the ethanol-rich vapor is allowed to condense to form a sufficiently pure alcohol. Distillation is an energy consuming process. The use of ethanol in petrofuels to increase the volume of fuel increases energy supply. By contrast, the need to consume energy to produce ethanol decreases energy supply. A favorable energy balance is achieved if the energy required to produce ethanol is less than the energy provided by the use of ethanol.

The adoption and widespread use of biofuels is hampered by cost when compared to the cost of using fossil fuels. One way to improve the economics of biofuels is to increase the cost of fossil fuel consumption. An increase in cost can be justified on environmental grounds by imposing a cost on the emission of greenhouse gases such as carbon dioxide. The cost can be imposed as a tax on the amount of greenhouse gases that are emitted by fossil fuel combustion, or by imposing a limit on the amount of greenhouse gases emitted.

Point to Ponder: Should farm land be used for food or fuel?
In recent years, biofuels have become increasingly popular as a means of extending the life of the world's oil supply. These fuel additives are renewable and reduce the amount of oil that must be used for transportation. This is particularly important in countries like the United States where there is a large transportation sector and a dwindling supply of oil. Still, biofuels have an additional cost that must be considered.

Some of the biomass feedstock for biofuels comes from organic byproducts. These scraps do not satisfy the demand for biomass in this growing industry. Furthermore, the United States government is an example of a government that subsidizes biofuels to encourage farmers to use their land for biomass rather than food stuffs. As a consequence, this creates a conflict between two choices: use land to grow food, or use land to grow feedstock for biofuels.

Societies must recognize that land area that is suitable for growing biofuels and food is limited. The designation of land for one product can reduce the amount of land available for growing the other product. Governments seeking to increase energy supplies must balance their need for biofuels with their need for food. This concept is an example of opportunity cost, which is discussed further in Chapter 12.

9.1.4 Biopower

Biomass has historically been used to provide heat for cooking and comfort. Today, biomass fuels can be used to generate electricity and produce natural gas. The power obtained from biomass, called biopower, can be provided in scales ranging from single-family homes to small cities. Biopower is typically supplied by one of four different classes of systems: direct-fired, cofired, gasification, and modular systems.

A direct-fired system is similar to a fossil fuel fired power plant. High-pressure steam for driving a turbine in a generator is obtained by burning biomass fuel in a boiler. Biopower boilers presently have a smaller capacity than coal-fired plants. Biomass boilers provide energy in the range of 20 to 50 megawatts, while coal-fired plants provide energy in the range of 100 to 1500 megawatts. The technology exists for generating steam from biomass with an efficiency of over 40%, but actual plant efficiencies tend to be on the order of 20% to 25%.

Biomass may be combined with coal and burned in an existing coal-fired power plant. The process is called cofiring and is considered desirable because the combustion of the biomass-coal mixture is cleaner than burning only coal. Biomass combustion emits smaller amounts of pollutants, such as sulfur dioxide and nitrogen oxides. The efficiency of the cofiring system is comparable to the coal-fired power plant efficiency when the boiler is adjusted for optimum performance. Existing coal-fired power plants can be converted to cofiring plants without major modifications. The efficiency of transforming biomass energy to electricity is comparable to the efficiency of a coal-fired power plant and is in the range of 33% to 37%. The use of biomass that is less expensive than coal can reduce the cost of operating a coal-fired power plant.

Biomass gasification converts solid biomass to flammable gas. Cleaning and filtering can remove undesirable compounds in the resulting biogas. The gas produced by gasification can be burned in combined-cycle power generation systems. The combined-cycle system combines gas turbines and steam turbines to produce electricity with efficiency as high as 60%.

Modular systems use the technologies described above, but on a scale that is suitable for applications that demand less energy. A modular system can be used to supply electricity to villages, rural sites, and small industry.

Point to Ponder: Is biomass a desirable alternative energy source?

Figure 9-5 shows a car that uses 100% vegetable oil as its fuel. At the time the photo was taken, a gallon of gasoline in Honolulu, Hawaii cost approximately one third more than a gallon of gasoline in Denver, Colorado. Increases in the price of gasoline stimulate the search for alternative transportation fuels. Vegetable oil is an example of a biomass alternative energy source.

Biomass provides energy by generating heat using combustion. One byproduct of the combustion process is carbon dioxide, a greenhouse gas. The benefits that biomass is renewable and can serve as an energy source are offset by the production of carbon dioxide.

Figure 9-5. Car Powered by 100% Vegetable Oil, Waikiki, Oahu (Fanchi, 2004)

9.2 Case Study: Biofuels in Brazil

Brazil is South America's leader in biofuel production and is second in the world only to the United States. Brazil began its ethanol production program in the early 1980s when the country was facing a major debt crisis because of the amount of oil it was importing [Goldemberg, 2008]. Brazilian leaders decided to take advantage of their consistently warm temperatures and tropical climate to produce large quantities of sugarcane to use as biomass for biofuels. Approximately half of Brazil's 2008 sugarcane crop is used for biofuel production.

When it became clear that Brazil must find a way to reduce its need for oil imports, the government began a large incentive program to encourage the development of biofuel crops and production technologies. Almost one-third of the financing required to develop the technology was invested between 1980 and 1985, and over the following two and a half decades the technology advanced to the point that government incentives were reduced to less than half of what they were originally. In 1980, the Brazilian government subsidized ethanol production at US$700 per 1000 liters (US$2.65 per gallon). By 2004, the subsidy had fallen to US$200 per 1000 liters (US$0.76 per gallon) as the industry approached self-sufficiency. This is a result of the increasing cost of oil, advances in technology and improvements in economies of scale.

Anhydrous ethanol, which is 99.6% ethanol and 0.4% water, is approximately 67% as efficient as gasoline. Hydrous ethanol (about 95.5% ethanol and 4.5% water) has a slightly lower efficiency. Both of these fuels have higher motor octane numbers than gasoline, however, meaning they can be used in a motor with a higher compression ratio and are thus 15% more efficient than gasoline in that sense. Overall, approximately 20% more ethanol than gasoline would typically be required per kilometer driven. The Brazilian government incentivized the use of ethanol as a stand-alone fuel in the 1980s, guaranteeing that the pump price of hydrous alcohol would be equivalent to 64.5% that of gasoline. This incentive led to an explosion in the popularity of ethanol-only cars and demand for ethanol-only fuel. Brazil experienced a shortage of ethanol-only fuel in the early 1990s which motivated a shift away from ethanol-only vehicles.

In 2003, Brazil introduced a flex-fuel motor that revolutionized the use of ethanol fuels. The flex-fuel motor is designed to use sensors to detect the ethanol level of the fuel in the tank. The motor then automatically adjusts the way it functions to reach maximum efficiency based on the fuel. This allowed for ethanol-gasoline mixtures of varying levels to be produced based on the best economics at the time and still be able to serve a large volume of motorists. While ethanol-only engines may be an option in the future, particularly if oil supplies are exhausted, the volume of ethanol needed to meet potential demand is not available.

The success of the flex-fuel motor implies that the demand for ethanol in Brazil can increase dramatically in the future. Brazil currently dedicates approximately 10% of its total farmland to sugarcane, and less than half of that is used for ethanol production. Brazil still believes that significant advances can be made in the efficiency of converting sugarcane and bagasse to ethanol. Demand for ethanol could quickly grow elsewhere in the world as other nations seek to emulate Brazil's advanced ethanol production, especially those countries with the capability to support large sugar-cane fields.

9.3 SYNFUELS

Synthetic fuels, or synfuels, are fossil fuel substitutes created by chemical reactions using such basic resources as coal or biomass. There are several ways to convert biomass into synfuels. Oils produced by plants such as rapeseed (canola), sunflowers and soybeans can be extracted and refined into a synthetic diesel fuel that can be burned in diesel engines. Thermal pyrolysis and a series of catalytic reactions can convert the hydrocarbons in wood and municipal wastes into a synthetic gasoline.

Society can extend the availability of fossil fuels such as natural gas and oil by substituting synthetic fuels for conventional fossil fuels. Here we consider coal gasification, biomass conversion, gas-to-liquids conversion, and algae oil.

9.3.1 Coal Gasification

Large coal molecules are converted to gaseous fuels in coal gasification. Coal is thermally decomposed at temperatures on the order of 600° Centigrade (620° Fahrenheit) to 800° Centigrade (980° Fahrenheit). The products of decomposition are methane and a carbon rich char. If steam is present, the char will react with the steam in the reaction to form carbon monoxide and hydrogen. The carbon monoxide-hydrogen mixture is called synthesis gas, or syngas. Carbon monoxide can react with steam in the reaction to form carbon dioxide and hydrogen. If the carbon dioxide is removed, the hydrogen-enriched mixture will react with carbon monoxide to produce methane and water vapor. The coal gasification process can be used to synthesize methane from coal. Methane is easier to transport than coal.

9.3.2 Biomass Conversion

Biomass, particularly biological waste, appears to be a plentiful source of methane. Methane production from biomass requires a conversion process. The simple methane digester shown in Figure 9-6 is a biomass conversion system that converts biological feed such as dung to methane. The screw agitator mixes the liquid slurry containing the feed. The mixing action facilitates the release of methane from the decaying biomass as a biomass conversion process. Methane trapped by the metal dome is recovered through the gas outlet.

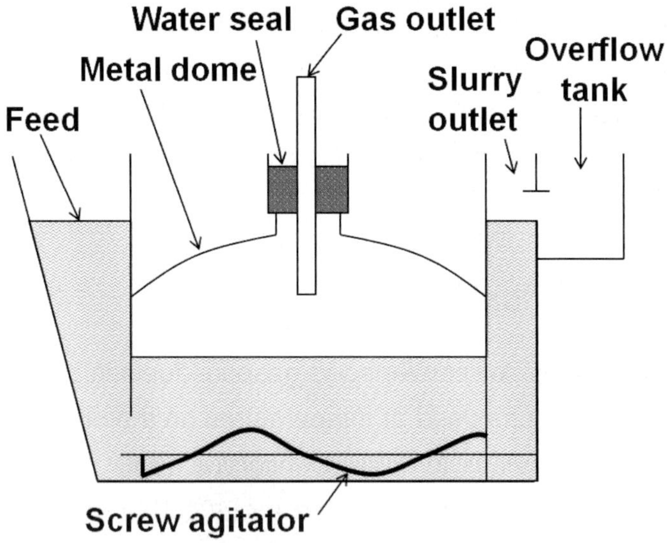

Figure 9-6. Methane Digester [after Cassedy and Grossman, 1998, page 298; and P.D. Dunn, Renewable Energies: Sources, Conversion and Application, 1986, P. Peregrinius, Ltd., London, United Kingdom]

Fermentation is another example of a biomass conversion process. Microorganisms are used in the fermentation process to convert fresh biological material into simple hydrocarbons or hydrogen. We illustrate fermentation processes by describing the anaerobic digestion process, a process that is well suited for producing methane from biomass.

The anaerobic digestion process proceeds in three stages. In the first stage, the complex biomass is decomposed by the first set of microorganisms. The decomposition of cellulosic material into the simple sugar glucose occurs in the presence of enzymes provided by the microorganisms. Stage one does not require an anaerobic (oxygen-free) environment. In the second stage, hydrogen atoms are removed in a dehydrogenation process that requires acidophilic (acid-forming) bacteria. In the third stage, a mixture of carbon dioxide and methane called biogas is produced from the acetic acid produced in stage two. The third stage requires the presence of anaerobic bacteria known as methanogenic bacteria in an oxygen-free environment.

Biomass conversion turns a waste product into a useful commodity. One difficulty with biomass conversion is the potential impact on the ecology of the region. For example, excessive use of dung and crop residues for fuel instead of fertilizer can deprive the soil of essential nutrients that are needed for future crops. Furthermore, the demand for large quantities of biomass requires the commitment of large areas of land and large volumes of water. Resource management must consider these issues.

9.3.3 Gas-to-Liquid Conversion

Synthetic liquid hydrocarbon fuels can be produced from natural gas by a gas-to-liquids (GTL) conversion process. Natural gas is composed of organic molecules such as methane and ethane which have a small number of carbon atoms. For example, methane has one carbon atom and ethane has two carbon atoms. The GTL process converts relatively small organic molecules into molecules with longer chains of carbon atoms. The primary product of the GTL process is a low sulfur, low aromatic diesel fuel. Aromatic molecules have multiple bonds between some of the carbon atoms and one or more carbon rings. Benzene, a carcinogen, is an example of an aromatic molecule.

9.3.4 Algae Oil

Algae are a plentiful biomass that can be processed to synthesize oil and gas. The synthetic process begins with the selection of a strain of algae based on such factors as rate of growth, survivability in different environments, and ability to produce hydrocarbons. The growth of algae requires water, nutrients and access to sunlight. When exposed to sunlight, algae can reproduce quickly and use photosynthesis to convert water and carbon dioxide into sugar. Algae cells metabolize the sugar into oily lipids, an organic material that can be extracted from algae cells. The extracted oil is refined to produce diesel oil using a catalyst to replace oxygen with hydrogen. Further refinement can alter the length of carbon chains to produce gasoline and jet fuel from algae oil.

Point to Ponder: Are synfuels desirable alternative energy sources?

Like biomass and fossil fuels, synfuels provide energy by generating heat using combustion. Once again the combustion process produces the greenhouse gas carbon dioxide as a by-product.

Unlike biomass, coal gasification and gas-to-liquid conversion depend on the fossil fuels coal and natural gas. These synfuels are only renewable to the extent that coal and natural gas are renewable.

Natural gas from the conversion of biological materials is renewable, but a byproduct of natural gas combustion is carbon dioxide, an undesirable greenhouse gas. The rate of generation of natural gas from biomass conversion limits the amount of energy that can be provided by biomass and synfuels.

9.4 Environmental Impact of Combustible Materials

The fuel used to drive a power plant is called the primary fuel. Examples of primary fuels include oil, coal, natural gas, and uranium. Fossil fuels are the primary fuel for most power plants in the world today. Fossil fuel resources are finite, however, and power plants that burn fossil fuels emit greenhouse gases and other pollutants. The extraction of some types of fossil fuels, such as coal and tar sands, can leave the land scarred and in need of reclamation (Figure 9-7).

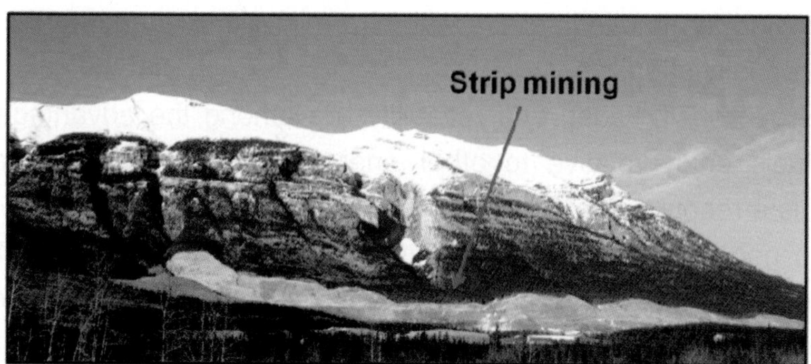

Figure 9-7. Scarred Land near Banff, Canada (Fanchi, 2002)

Fossil fuels are combustible fuels: energy is released from the fuel by combustion. Biomass and synfuels increase the supply of combustible materials. One important environmental consequence of using a combustible material is the production of greenhouse gases. The apparently obvious conclusion is to replace fossil fuels with renewable sources of energy that are environmentally benign. Before judging fossil fuels too harshly, however, it must be realized that every energy source has advantages and disadvantages. The environmental impact of some clean energy sources is summarized below:

- Hydroelectric facilities, notably dams, can flood vast areas of land. The flooded areas can displace people and wildlife, and impact the ecosystems of adjacent areas with consequences that may be difficult to predict. Dams can change the composition of river water downstream of the dam and can deprive land areas of a supply of silt for agricultural purposes. A dam on a river can prevent the upstream migration of certain species of fish, such as salmon.

- Geothermal power plants can emit toxic gases such as hydrogen sulfide or greenhouse gases such as carbon dioxide. The produced water from a geothermal reservoir will contain dissolved solids that can form solid precipitates when the temperature and pressure of the produced water changes.
- Solar power plants are relatively inefficient, have a relatively large footprint, and may be visually offensive to some people.
- Wind farms can interfere with bird migration patterns and may be visually offensive.

Energy density, cost, and reliability are among the advantages that fossil fuels enjoy relative to other energy sources. A reliable energy source is a resource that is available almost all the time. Some downtime may be necessary for facility maintenance or other operating reasons. The United States Department of Energy has defined a reliable energy source as an energy source that is available at least 95% of the time [DoE Geothermal, 2002]. The selection of an energy source should consider such factors as energy density, cost, reliability, social acceptability, and environmental impact.

9.4.1 Climate Change

Climate change refers to the change in a property of the atmosphere that lasts for a long period of time. Atmospheric properties include temperature and precipitation patterns. Global warming refers to the increase in atmospheric temperature over a period of decades. Measurements of ambient air temperature show a global warming effect that corresponds to an increase in the average temperature of the earth's atmosphere (Figure 9-9). The average temperature at the earth's surface is approximately 14° Celsius (57° Fahrenheit), and typically varies from -53°

Celsius (-64° Fahrenheit) to 47° Celsius (116° Fahrenheit) [Sørensen, 2000, page 26]. Figure 9-9 shows that average global temperature has been increasing since around 1910, although average global temperature declined in recent years (between 2005 and 2008). The baseline temperature change (-0.0 on the vertical axis) in the figure is the mean temperature from 1961-1990.

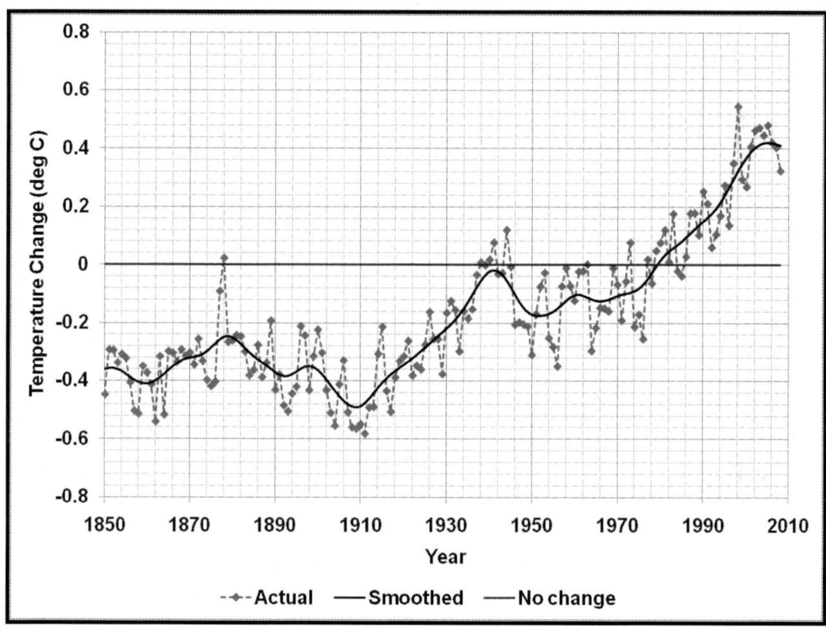

Figure 9-9. Historical Temperature of the Earth's Atmosphere [data from CRU website, 2010]

The increase in temperature was coincident with the historical use of fossil fuels beginning in the 19th century. Figure 9-10 shows that the increase in temperature is also coincident with an increase of CO_2 composition in the atmosphere, and an increase in global population. The use of fuels with energy densities larger than biomass energy density made it possible for societies to support larger populations. The increase in population was accompanied by decrease in forests as wood was used for fuel and construction. Scientists Albert A. Bartlett and Paul B. Weisz argued that growth in the world's population is a major cause of societal problems [Bartlett, 2004; Weisz, 2004], and more attention should be focused on stopping and reversing population growth.

The term global warming was used to alert people that continuing use of fossil fuels may be leading to environmentally significant changes in the temperature of the atmosphere. The decade from 2000 to 2009 showed that the temperature of the atmosphere is declining, and the term climate change became more fashionable because it referred to the entire set of atmospheric properties that have been changing in conjunction with societal use of fossil fuels.

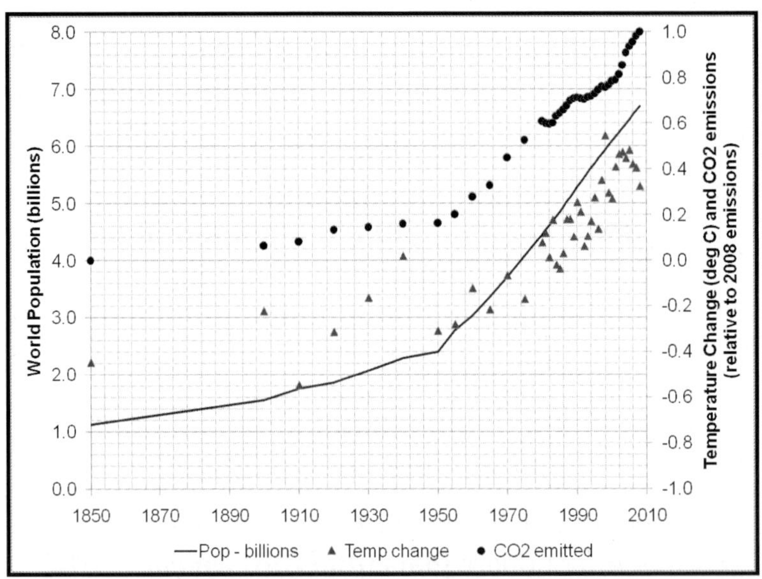

Figure 9-10. Comparison of Historical Population, Temperature, and CO_2 Emissions [US Census, 2010; US EIA website, 2010]

The increase in atmospheric temperature is associated with the combustion of fossil fuels. When a carbon-based fuel burns, carbon can react with oxygen and nitrogen in the atmosphere to produce carbon dioxide, carbon monoxide, and nitrogen oxides (NOx). The combustion byproducts, including water vapor, are emitted into the atmosphere in gaseous form. Some of the gaseous byproducts are called greenhouse gases because the gases capture energy from sunlight that is reflected by the earth's surface. Greenhouse gases include carbon dioxide, methane, and nitrous oxide, as well as other gases such as volatile organic compounds and hydrofluorocarbons. Figure 9-11 shows the increase in greenhouse gas concentrations since the 1800s.

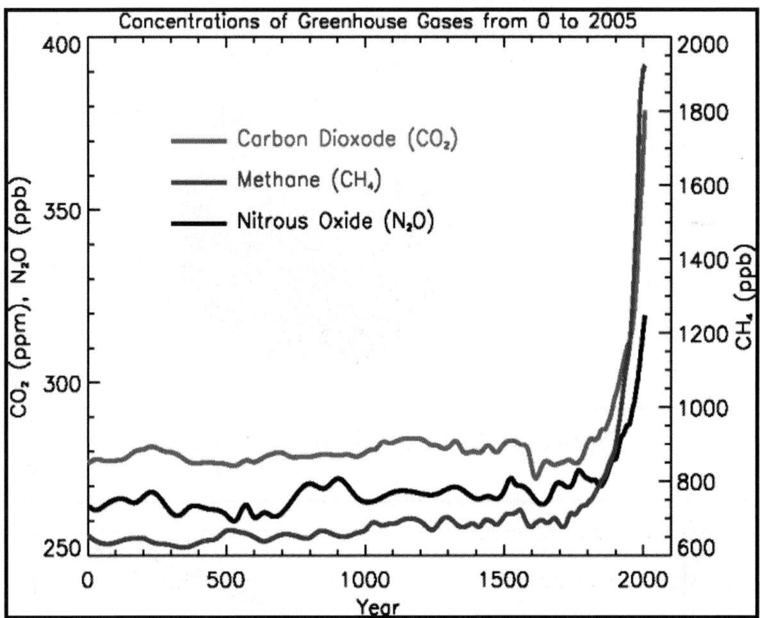

Figure 9-11. Greenhouse Gas Concentrations in the Atmosphere [IPCC, 2007]

The Greenhouse Effect is illustrated in Figure 9-12. Some of the incident solar radiation from the sun is absorbed by the earth, some is reflected into space, and some is captured by chemicals in the atmosphere and reradiated as infrared radiation (heat). The reradiated energy would have escaped the earth as reflected sunlight if greenhouse gases were not present in the atmosphere.

Carbon dioxide (CO_2) is approximately 83% by mass of the greenhouse gases emitted by the United States. As of November 2009, atmospheric CO_2 concentration was 388 parts per million [NASA website, http://climate.nasa.gov/keyIndicators/]. For comparison, pre-industrial atmospheric CO_2 concentration was on the order of 288 parts per million.

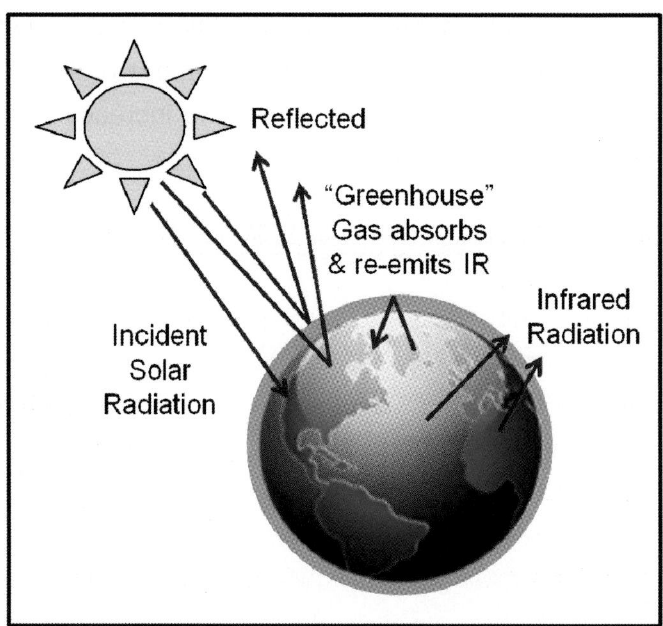

Figure 9-12. The Greenhouse Effect

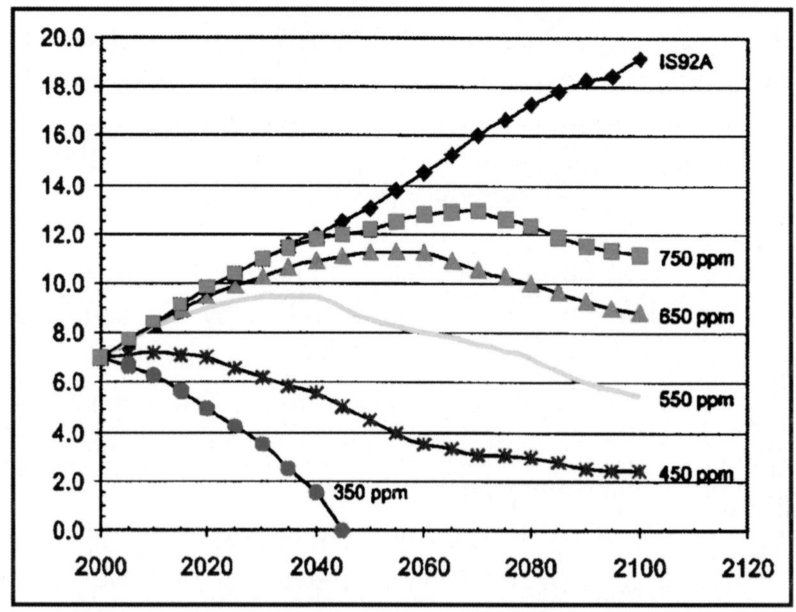

Figure 9-13. Projection of CO_2 Concentration in the Earth's Atmosphere [Wigley, et al., 1996]

T.M.L. Wigley, R. Richels and J.A. Edmonds [1996] projected ambient CO_2 concentration through the 21st century. Their results are shown in Figure 9-13. The concentration of CO_2 that would establish an acceptable energy balance is considered to be 550 parts per million. To achieve the acceptable concentration of CO_2 through the next century, society would have to reduce the volume of greenhouse gases entering the atmosphere.

Proponents of global climate change consider a number of indicators along with carbon dioxide levels. Two other key factors are changes in global sea level and changes in Arctic sea ice. These are related since increasing global temperature is expected to cause melting of Arctic sea ice and lead to a rise in global sea level. Taken together, the evidence that human activity is causing the climate to change more than it would naturally change has motivated international attempts to regulate greenhouse gas emissions.

The Kyoto Protocol is an international treaty that was negotiated in Kyoto, Japan in 1997 to establish limits on the amount of greenhouse gases a country can release into the atmosphere. The Kyoto Protocol was not accepted by the United States. Some countries believed the greenhouse gas emission limits were so low that emission limits would adversely impact national economies without solving the problem of global warming. The Kyoto Protocol placed more controls on greenhouse gas emissions from developed countries than it did on emissions from developing nations like India and China, which have rapidly expanding fossil-fuel based economies and approximately a third of the world's population. Another attempt to establish an international framework to regulate greenhouse gas emissions was held in Copenhagen in December 2009. The Copenhagen conference resulted in a non-binding statement of intent to control emissions.

Research is underway to develop the technology needed to capture and store greenhouse gases. One storage process called geologic sequestration is designed to store greenhouse gases in geologic formations. Geologic sequestration is an economically viable means of mitigating the increase in greenhouse gas concentration in the atmosphere.

Point to Ponder: Is geologic sequestration a desirable policy?

It would seem that geologic sequestration would be a desirable policy for a society that is dependent on fossil fuels and is trying to minimize carbon dioxide emissions. On the other hand, geologic sequestration can help a society justify continued use of fossil fuels and reduce its motivation to develop an energy infrastructure that supports sustainable energy sources rather than fossil fuels. From this perspective, geologic sequestration is an undesirable policy.

9.5 Activities

True-False

Specify if each of the following statements is True (T) or False (F).

1. Photosynthesis occurs in plants and bacteria.
2. Biological conversion of solar energy relies on photosynthesis.
3. Each stage of anaerobic digestion requires oxygen.
4. Most of the energy that is radiated by the sun reaches the surface of the earth.
5. Biomass includes matter that was recently formed as a result of photosynthesis.
6. Biomass gasification converts solid biomass to flammable gas.
7. A methane digester produces methane from a biological feedstock.
8. An anaerobic process requires oxygen.
9. Gas is generated in a landfill by the decay of organic material.
10. Synthetic diesel can be obtained from plants.
11. The Greenhouse Effect is the absorption of sunlight and emission of infrared radiation.

Questions

1. What are the 3 stages of anaerobic microbial degradation?
2. A landfill generates 4 million cubic feet of gas per day. What is the cash flow (in US$ per day) of the gas if the price of gas is US$5 per thousand cubic feet of gas?

3. What is the product X in the combustion reaction $C+O_2 \rightarrow X$?
4. What impact does carbon dioxide have on the environment? You should consider the atmosphere and the hydrosphere.
5. How is ethanol used in transportation?
6. Match the definition to the term by writing the letter of the definition in the column to the left of the term.

	Term		**Definition**
	Synfuels	A	Matter that was recently formed as a result of photosynthesis
	Methane digestor	B	Fossil fuel substitutes created by chemical reactions using basic resources, e.g. coal
	Geological sequestration	C	Storage of greenhouse gases in subsurface formations (e.g. reservoirs)
	Biomass	D	System that converts biological feed into methane

7. What is geologic carbon sequestration?
8. What is the Kyoto Protocol?
9. Did the United States legally agree to abide by the Kyoto Protocol?
10. Match the adverse environmental impact to the clean energy technology by writing the letter of the adverse environmental impact in the column to the left of the clean energy technology.

	Clean Energy Technology		**Adverse Environmental Impact**
	Dam	A	Large footprint due to heliostat field
	Geothermal power plant	B	Interference with bird migration patterns
	Solar tower	C	Emission of toxic gases
	Wind farm	D	Flood vast areas of land

CHAPTER 10

ENERGY CARRIER, ENERGY STORAGE AND HYBRID ENERGY SYSTEMS

Energy sources can be classified as renewable and non-renewable. Some energy sources like coal, oil and gas are relatively easy to transport and store. Some energy sources like wind and solar are intermittent energy sources that require backup to meet demand.

Demand for electric power is the load on a utility. Utilities need to have power plants that can meet three types of loads: base load, intermediate or cycling load, and peak load. The base load is the minimum baseline demand that must be met in a 24-hour day. Intermediate load is the demand that is required for several hours each day and tends to increase or decrease slowly. Peak load is the maximum demand that must be met in a 24-hour day. Today, power plants that rely on fossil fuels and nuclear energy provide base load power, and wind farms and solar electric generating facilities provide supplemental power. If we want to rely on renewable energy in the future for all of our power needs, it will be necessary to find a way to provide power when intermittent energy sources are offline.

In addition, energy from nuclear, wind, and solar energy sources use electricity as an energy carrier, that is, an agent for distributing energy. Electricity has proven to be difficult to store in sufficient quantities for many commercial applications. An alternative energy carrier that has more useful transport and storage properties than electricity is hydrogen. To optimize the use of nuclear, wind and solar energies, which are plentiful sources of energy, we need to resolve the energy transport and storage issues. The following possible solutions are discussed here: hydrogen as an energy carrier, large scale energy storage systems, and hybrid energy systems.

10.1 Hydrogen

Hydrogen is found almost everywhere on the surface of the earth as a component of water. Each water molecule in the mountain reservoir in Summit County, Colorado (Figure 10-1) contains two atoms of hydrogen and one atom of oxygen. Hydrogen has many commercial uses, including production of ammonia for use in fertilizers, methanol production, hydrochloric acid production, and use as a rocket fuel. Liquid hydrogen is used in cryogenics and superconductivity. Hydrogen is important to us because it can be used as a fuel that is relatively easy to store and transport.

Figure 10-1. Hydrogen in its Most Common Form on Earth (Fanchi, 2000)

Hydrogen can be used as a fuel for a modified internal combustion engine or in a fuel cell. Fuel cells are electrochemical devices that directly convert hydrogen, or hydrogen-rich fuels, into electricity using a chemical process. Fuel cells do not need recharging or replacing and can produce electricity as long as they are supplied with hydrogen and oxygen. Hydrogen can be produced by the electrolysis of water, which uses electrical energy to split water into its constituent elements. Electrolysis is a net energy consuming process in which more energy is consumed than produced. Hydrogen can also be produced from certain types of algae. Hydrogen is not considered an energy source because more energy is required to produce hydrogen than can be obtained from hydrogen. Hydrogen is considered a carrier of energy because it can be used as a fuel to provide energy. The properties of hydrogen make it a source of hope for a sustainable, global energy system. Here we first describe the properties of hydrogen before discussing the production of hydrogen and its use as the primary fuel in fuel cells.

10.1.1 Properties of Hydrogen

Hydrogen is the first element in the Periodic Table. The nucleus of hydrogen is the proton and hydrogen has only one electron. At ambient conditions on Earth, hydrogen is a colorless, odorless, tasteless, and non-toxic gas of diatomic molecules, that is, a molecule with two atoms with the symbol H_2. Selected physical properties of hydrogen, methane and gasoline are shown in Table 10-1. The properties of gasoline are illustrative since there are many grades of gasoline.

A kilogram (2.2 pounds mass) of hydrogen, in either the gas or liquid state, has a greater energy density than the most widely used fuels today, such as oil and coal. The volumetric energy density of hydrogen is much less than gasoline however, which means that a larger volume of hydrogen is needed to provide the same energy as gasoline, or hydrogen must be compressed and stored under pressure.

Table 10-1
Selected Physical Properties of Hydrogen, Methane, and Gasoline [a]

Property	Hydrogen (gas)	Methane (gas)	Gasoline (liquid)
Molecular Weight (grams/moles) [a]	2.016	16.04	~110
Mass Density (kilograms/meters^3) [a,b]	0.09	0.72	720-780
Energy Density (megajoules/kilograms)	120 [a]	53 [c,d]	46 [a,c]
Volumetric Energy Density (megajoules/meters^3) [a]	11 [a]	38 [c,d]	35,000 [a,c]
Higher Heating Value (megajoules/kilograms) [a]	142.0	55.5	47.3
Lower Heating Value (megajoules/kilograms) [a]	120.0	50.0	44.0

a. Ogden [2002, Box 2, page 71]
b. at 1 atm and 0 °C
c. Hayden [2001, page 183]
d. Ramage and Scurlock [1996, Box 4.8, page 152]

Point to Ponder: Is hydrogen the perfect fuel of the future?
Hydrogen is considered a clean, reliable fuel once it is produced because the combustion of hydrogen produces water vapor. Hydrogen can react with oxygen to form water in an exothermic combustion reaction. Hydrogen combustion does not emit toxic greenhouse gases like carbon monoxide or carbon dioxide. When hydrogen is burned in air, it does produce traces of nitrogen oxides, which are pollutants. A particularly costly problem with hydrogen is that hydrogen is not readily available as free

hydrogen; hydrogen is most readily available in combination with other elements, such as oxygen in water. Consequently, elemental hydrogen must be freed before it can be used as a fuel. The production of hydrogen is discussed below.

10.1.2 Hydrogen Production

Hydrogen can be produced by electrolysis. Electrolysis is the non-spontaneous splitting of a substance into its constituent parts by supplying electrical energy. In the case of water, electrolysis would decompose the water molecule into its constituent elements (hydrogen and oxygen) by the addition of electrical energy.

It is difficult to electrolyze very pure water because there are few ions present to flow as an electrical current. Electrolyzing an aqueous salt solution enhances the production of hydrogen by the electrolysis of water. An aqueous salt solution is a mixture of ions and water. The addition of a small amount of a non-reacting salt such as sodium sulfate accelerates the electrolytic process. The salt is called an electrolyte and provides ions that can flow as a current.

We illustrate the electrolysis process in Figure 10-2 using the electrolysis of water-salt solution. The salt in our example is table salt. Chemically, table salt is sodium chloride and is made up of one atom of sodium (Na) and one atom of chlorine (Cl). Sodium chloride (NaCl) separates into electrically charged Na^+ and Cl^- ions when dissolved in water. The ions respond to an electrical potential difference and make the water-salt solution electrically conductive.

The electrolytic cell in Figure 10-2 is a cell with two electrodes: the anode and the cathode. The voltage source (such as a battery) acts like an electron pump that pulls electrons from the anode and pushes them to the cathode. In the figure, a negatively charged chlorine ion with one extra electron loses its excess electron at the anode and combines with another chlorine atom to form the chlorine molecule (Cl_2). Electrons flow from the anode through an external circuit to the cathode where positively

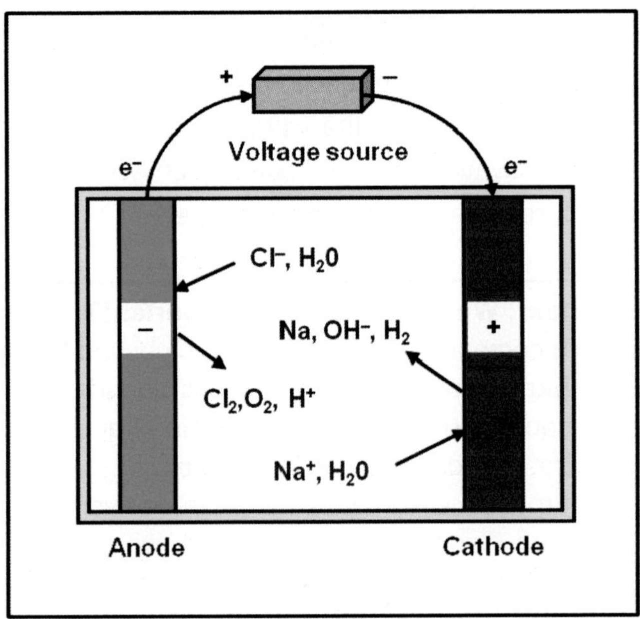

Figure 10-2. Electrolysis of Water-Salt (NaCl) Solution

charged sodium ions combine with the electrons and become electrically neutral atoms of sodium.

Electrolysis is a reduction-oxidation (redox) reaction. Oxidation refers to the loss of electrons and occurs when electrons are released by a chemical species. Reduction refers to the gain of electrons and occurs when electrons are accepted by a chemical species. Oxidation occurs at the anode, and reduction occurs at the cathode. The electrolysis reaction can be viewed as two half-reactions: the oxidation half-reaction at the anode, and the reduction half-reaction at the cathode. A reducing agent or reductant is a chemical species that donates electrons in a reaction, and an oxidizing agent or oxidant is a chemical species that accepts electrons in a reaction.

The overall reaction for electrolysis of a water-salt solution in an electrolytic cell shows that hydrogen gas is produced at the cathode of the electrolytic cell and oxygen gas is produced at the anode when enough electrical energy is supplied. Four hydrogen ions combine with four hydroxyl ions to form four molecules of water. Electrical energy is supplied in the form of an electrical voltage and must be greater than the threshold energy (activation energy) of the reaction. The voltage needed to electrolyze water must be greater than the voltage that we would predict if we did not consider activation energy; consequently, the actual voltage needed to electrolyze water is called overvoltage.

Point to Ponder: Why is electrolysis important?
Hydrogen is a component of water. The distribution and amount of water on Earth implies that we have an abundant source of hydrogen for energy use if we can separate hydrogen from the water molecule, collect it and use it as a fuel.

For our purposes, electrolysis can produce hydrogen for use as a fuel. When electrolysis is completed, we have produced molecular hydrogen and molecular oxygen from water molecules. Hydrogen is a convenient fuel because it is relatively easy to store and it is portable. Hydrogen is not a source of energy, however, because the energy to dissociate water into hydrogen and oxygen is greater than the energy provided by the produced hydrogen. That is why electrolysis is considered a net energy consuming process and hydrogen is considered an energy carrier.

10.1.3 Thermal Decomposition and Gasification

Electrolysis is one technique for producing hydrogen. Two other techniques are thermal decomposition and gasification. The process of decomposing the water molecule at high temperatures is called thermal decomposition. Heat in the form of steam can be used to reform hydrocarbons and produce hydrogen.

Steam reforming exposes a hydrocarbon such as methane, natural gas, or gasoline to steam at high temperature (850° Centigrade) and high pressure (2.5×10^6 Pascal) [Sørensen, 2000, pages 572-573]. Steam reforming reactions produce hydrogen. One steam reforming reaction uses hydrocarbons, which we have seen have limited long-term potential because hydrocarbons are not considered a renewable source of energy. We can produce a hydrocarbon like natural gas using a technique such as methane digestion, but the rate of production is relatively slow compared to the current rate of consumption. A process such as absorption or membrane separation can be used to remove the carbon dioxide byproduct. The production of carbon dioxide, a greenhouse gas, is an undesirable characteristic of steam reforming.

Some bacteria can produce hydrogen from biomass by fermentation or high-temperature gasification. The gasification process is similar to coal gasification.

Point to Ponder: Where do we get the energy we need to produce hydrogen?

One of the problems facing society is how to produce hydrogen using an environmentally acceptable process. Three possible sources of energy for producing hydrogen in the long-term are wind energy, nuclear fusion, and solar energy. We have already seen that wind energy and solar energy are not always produced when there is a demand for energy. One way to enhance the usefulness of wind and solar energy is to use excess wind and solar energy to produce hydrogen during periods when the wind is blowing and the sun is shining. The produced hydrogen could then be used as a source of energy when wind energy and solar energy are not being generated.

J.H. Ausubel [2000] has suggested that the potential of nuclear energy will be realized when nuclear energy can be used as a source of electricity and high-temperature heat for splitting water into its constituent parts. In this scenario, nuclear energy could be generated at sites with excess energy generating capacity and then used to produce hydrogen. The hydrogen would then fulfill its role as a carrier of energy.

10.1.4 Hydrogen and Fuel Cells

Hydrogen is a carrier of energy that can be transported in the liquid or gaseous state by pipeline or in cylinders. Once produced and distributed, hydrogen can be used as a fuel for a modified internal combustion engine or as the fuel in a fuel cell. Fuel cells are electrochemical devices that directly convert hydrogen, or hydrogen-rich fuels, into electricity using a chemical rather than a combustion process.

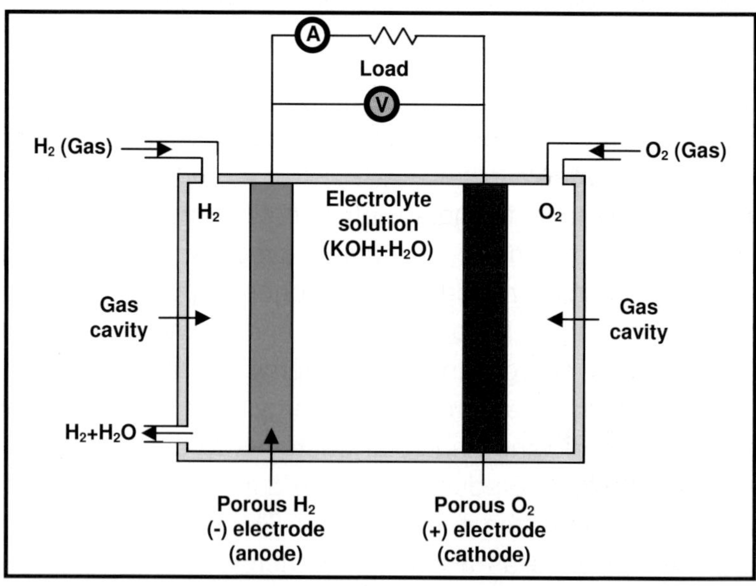

Figure 10-3. Schematic of a Fuel Cell [after Cassedy and Grossman, 1998, page 419; and Crowe, 1973]

A fuel cell consists of an electrolyte sandwiched between an anode and a cathode (Figure 10-3). The electrolyte is a substance that contains free ions so that the substance is electrically conductive. A water-salt solution is a common electrolyte because salt is composed of molecules that can separate into positively charged ions and negatively charged ions when dissolved in water. The presence of ions increases the electrical conductivity of the water-salt solution. The electrolyte in Figure 10-3 is a mixture of potassium hydroxide and water.

The electrolyte solution is maintained at a lower pressure than the gas cavities on either side of the porous electrodes. The pressure gradient facilitates the separation of hydrogen and oxygen molecules. The type of electrolyte in a fuel cell distinguishes that fuel cell from other types of fuel cells. The load in the figure is a circuit with amperage *A* and voltage *V*. Hydrogen is fed to the anode (negative electrode) and oxygen is fed to the cathode (positive electrode). When activated by a catalyst, hydrogen atoms dissociate (or separate) into protons and electrons. The charged protons and electrons take different paths to the cathode. The electrons go through the external circuit and provide an electrical current, while the protons migrate through the electrolyte to the cathode. Once at the cathode, the protons combine with electrons and oxygen to produce water and heat. The heat is waste heat unless it can be captured for use in a co-generation process. Produced water must be removed from the fuel cell.

The fuel cell in Figure 10-3 is a proton exchange membrane (PEM) fuel cell. The PEM fuel cell depends on the movement of protons (hydrogen nuclei) through porous electrodes.

Fuel cells produce clean energy in the form of electricity and heat from hydrogen. Fuel cells do not need recharging or replacing and can produce electricity as long as they are supplied with hydrogen and oxygen. Hydrogen and oxygen are fuels that must be accessible from some suitable storage location.

10.2 The Hydrogen Economy

The historical trend toward decarbonization reflects the contention by many energy forecasters that hydrogen will be the fuel of choice in the future. These forecasters believe that power plants and motor vehicles will run on hydrogen. The economies that emerge will depend on hydrogen and are called hydrogen economies.

The concept of a hydrogen economy is not new. The use of hydrogen as a significant fuel source driving a national economy was first explored in the middle of the 20th century as a complement to the adoption of large scale nuclear electric generating capacity. Concerns about global climate change and the desire to achieve sustainable development have renewed interest in hydrogen as a fuel.

A future that depends on hydrogen is not inevitable because many challenges remain. Hydrogen economies will require the development of improved technologies for producing, storing, transporting, and consuming hydrogen. We have already discussed some of the challenges involved in the production of hydrogen. Another example of a technological challenge that must be overcome in a transition to a hydrogen economy is the ability to store hydrogen.

Hydrogen can be stored in the liquid or gaseous state, but it must be compressed to high pressures or liquefied to achieve reasonable storage volumes because of the low density of the diatomic hydrogen molecule. The energy content of hydrogen gas is less than the energy contained in methane at the same temperature and pressure. The volumetric energy density of liquid hydrogen is approximately 8700 megajoules/meter^3. This is about one third the volumetric energy density of gasoline. The relatively low volumetric energy density of hydrogen creates a storage problem if we want to store hydrogen compactly in vehicles.

Hydrogen can be stored effectively in the form of solid metal hydrides. A metal hydride is a metal that absorbs hydrogen. The hydrogen is absorbed into the spaces, or interstices, between atoms in the metal. According to M. Silberberg [1996, page 246], metals such as palladium and niobium "can absorb 1000 times their volume of H_2 gas, retain it under normal conditions, and release it at high temperatures." This form of storage may be desirable for use in hydrogen-powered vehicles. A model of a car that uses hydrogen fuel cells is shown in Figure 10-4.

Hydrogen can be hazardous to handle. A spectacular demonstration of this fact was the destruction of the German zeppelin *Hindenburg*. The *Hindenburg* used hydrogen for buoyancy. In 1937, the *Hindenburg* burst into flames while attempting a mooring in Lakehurst, New Jersey. At the time, people believed that the hydrogen in the *Hindenburg* was the cause of the explosion. Addison Bain, at the end of the 20th century, showed that the chemical coating on the outside of the zeppelin was the cause of the explosion. When the chemical coating ignited, the hydrogen began to burn. Today, lighter-than-air ships use less flammable gases such as helium.

Figure 10-4. Hydrogen Fuel Cell Car, Denver, Colorado Exhibit (Fanchi, 2003)

Hydrogen, natural gas and gasoline are flammable in air. The two major chemical components in air are nitrogen and oxygen. Hydrogen forms an explosive mixture with air when the concentration of hydrogen in air is in the range of 4% to 75% hydrogen. For comparison, natural gas is flammable in air when the concentration of natural gas in air is in the range of 5% to 15% natural gas. Furthermore, the energy to ignite hydrogen-air mixtures is approximately one-fifteenth the ignition energy for natural gas-air or gasoline-air mixtures. The lower ignition energy of hydrogen in air makes it possible to consider hydrogen a more dangerous fuel than natural gas. On the other hand, the low density of hydrogen allows hydrogen to dissipate more quickly into the atmosphere than a higher density gas such as methane. Thus, hydrogen leaks can dissipate more rapidly than natural gas leaks. Adding an odorant to the gas can enhance the detection of gas leaks.

The environmental acceptability of hydrogen fuel cells depends on how the hydrogen is produced. If a renewable energy source such as solar energy is used to generate the electricity needed for electrolysis, vehicles powered by hydrogen fuel cells would be relatively clean since hydrogen combustion emits water vapor. Unfortunately, hydrogen combustion in air also emits traces of nitrous oxide compounds. Nitrogen dioxide contributes to photochemical smog and can increase the severity of respiratory illnesses.

Point to Ponder: Would people be willing to use hydrogen in their everyday lives?

The common conception of hydrogen depends on a person's cultural background. People in the United States tend to think of the *Hindenburg* on fire when they think of hydrogen as a fuel. One of the tasks of an evolving hydrogen industry is to educate the public about the misconceptions associated with hydrogen.

One example of the education effort is to re-educate the public about the *Hindenburg* disaster. Hydrogen economy proponents have presented Bain's new explanation of the *Hindenburg* disaster at trade shows and public exhibitions. Videos of staged automobile accidents are another example of the education effort. The videos show how a hydrogen-fueled car would burn following an accident compared to a gasoline-fueled car. Hydrogen from a breach in the storage tank of a hydrogen-fueled car is less buoyant than air and would rise, while gasoline from a breach in the storage tank of a gasoline-fueled car would pool on the ground around the vehicle. The gasoline could ignite and engulf the vehicle, while the hydrogen would form a vertical flare of flame escaping from the breach. The videos show that hydrogen could be used as a relatively safe fuel in transportation. These examples show that proponents of the hydrogen economy believe that an education effort is needed to overcome social resistance to adopting hydrogen as a primary fuel.

10.3 Large Scale Energy Storage

Solar and wind energy are two renewable energy sources that hold great promise for replacing fossil fuels as the dominant energy sources in the future energy mix. The amount of wind and solar energy that can be converted to useful power depends on location, time of day, season, and weather conditions. Consequently, both solar and wind energy are intermittent sources of energy with limited reliability; they are not available all day long, every day of the year.

The problem of intermittent reliability is currently solved by using a more reliable source of energy, such as non-renewable fossil fuels or nuclear energy, to provide energy as needed. In this power configuration, solar and wind energy are used to supplement other forms of energy. Our ability to rely on solar and wind energy as baseline sources of energy depends on our ability to store solar and wind energy in forms that can be readily accessed. One way to optimize the performance of intermittent energy sources is to use large scale energy storage, which is the subject of this section.

10.3.1 Energy Storage Concepts

We saw in Chapter 1 that energy exists in many different forms, and that energy can be transformed from one form to another. The transformation property of energy is being used to store energy in a variety of forms, including chemical, biological, electrical, electrochemical, mechanical, and thermal. Each of these forms is introduced here.

Energy can be stored in chemical form as an energy carrier such as hydrogen or as a biofuel or synfuel. Chemical reactions like combustion release the stored chemical energy.

Energy can be stored in biological substances such as starch or glycogen. Starch is a carbohydrate that stores energy from photosynthesis. Glycogen stores energy for use in animal cells.

Energy can be stored in electrical form in devices such as capacitors and superconducting magnetic energy storage (SMES) systems. A capacitor stores electrical energy in the dielectric material sandwiched between two conductors. The SMES system stores energy in the magnetic field that is induced by the direct current in a superconducting coil. The coil has to be cooled to very low temperatures so that it can be superconducting. Superconductors are materials that offer no resistance to electron flow.

Dutch physicist Hans Kamerlingh Onnes (1853-1926) discovered the first superconductor in 1911. Onnes produced liquid helium by cooling helium to a temperature below 4.2° Kelvin (-269° Celsius). He then observed that the resistivity of mercury vanished when mercury was cooled by liquid helium. Superconducting materials have been developed that operate at higher temperatures than 4.2° Kelvin, but superconductors can still only operate at temperatures that are far below room temperature. Once the coil is cool enough to operate as a superconductor, the current does not encounter any resistance and the energy in the magnetic field induced by the current can be stored indefinitely. It is important to note that energy is needed to refrigerate the superconducting coil, so the storage of magnetic energy requires the consumption of energy for refrigeration.

Energy can be stored in electrochemical form in devices such as batteries and fuel cells. These devices use redox reactions to generate a flow of electrons (see Section 10.1).

Energy can be stored in mechanical form using a variety of mechanisms. One mechanical energy storage mechanism is compressed air energy storage (CAES). CAES uses off-peak energy from a power plant to inject gas into a high pressure reservoir. The gas is withdrawn from the reservoir to drive a gas turbine when additional power is needed.

Another mechanical energy storage mechanism is hydroelectric energy storage. The method uses off-peak energy from a hydroelectric power plant to pump water from a lower elevation to a storage pool at a higher elevation. The elevated water is used to drive a water turbine when additional power is needed.

Energy can be stored in thermal form using a variety of mechanisms. One thermal energy storage mechanism is molten salt. We discussed one application of molten salt thermal energy storage in Chapter 6. Molten salt was used to store solar energy. Solar energy can also be used to heat a saltwater pool. The pool is called a solar pond and can be used for applications such as heating and desalination.

Another thermal energy storage mechanism is a steam accumulator. Off-peak energy can be used to heat water, which is stored as hot water or steam under pressure in an insulated steel pressure tank. The hot water or steam is used to drive a turbine when additional power is needed. Steam accumulators are being used at the PS10 solar power tower plant near Seville, Spain.

The energy storage techniques discussed above are in various stages of development. Some of the techniques are not well suited for large scale energy storage, and others use too much energy or are too costly to be commercial. CAES and steam accumulators are suitable for large scale energy storage.

10.3.2 Compressed Air Energy Storage

Compressed Air Energy Storage (CAES) is an example of a large scale energy storage technology that has been implemented in Germany and Alabama. CAES is designed to make off-peak energy from primary power plants accessible during peak demand periods. Compressed air can be stored in solution-mined salt cavities, mined hard rock cavities, aquifers, geologic reservoirs, or storage tanks. The gas storage technique depends on location of the site, economics, commercial objectives and government regulations.

The Huntorf CAES facility in Germany and the McIntosh CAES facility in Alabama store gas in salt caverns. The CAES facility in Huntorf was built in 1978 and is the world's first CAES project. Gas was stored in the two salt caverns shown in Figure 10- 5. The power plant in the figure can generate 290 MW. Off-peak energy at Huntorf is used to pump air underground and compress it in a salt cavern. The compressed air is produced during periods of peak energy demand to drive a turbine and generate additional electrical power.

The McIntosh power plant is a gas-fired power plant with approximately 340 MW capacity. The CAES facility associated with the plant, designated McIntosh unit 1, came online in 1991. During peak hours when additional energy production capacity is needed, the CAES facility produces compressed air that is added to a natural gas mixture for combustion.

Figure 10-5. Huntorf CAES Facility

Currently, CAES systems have been limited to areas where power plants and salt formations coexist. There are many other areas where current or proposed wind farms, solar power plants, and nuclear power plants, among others, overlie permeable formations that have or could contain gas. The potential for widespread implementation of CAES technology is significant.

10.3.3 Renewable Energy and CAES

Intermittent sources of energy such as wind and solar energy can function as the primary energy source for a utility if they can reliably provide baseline electricity directly to the electrical grid. One way to store solar and wind energy is to convert these energy forms to other forms of energy. For example, excess solar energy from a solar power plant could be used to heat and compress a fluid that could be stored in a reservoir. The hot, compressed fluid could be extracted from the reservoir and used to generate electrical power when the amount of energy from the solar power plant declines. Similarly, excess wind energy from a wind farm could be used to compress a fluid that could be stored in a reservoir.

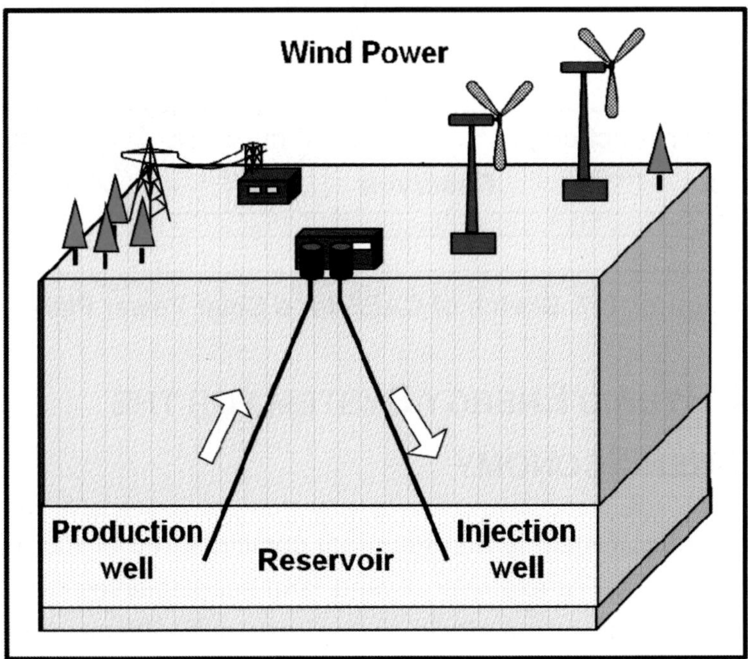

Figure 10-6. Sketch of CAES for a Wind Farm

A sketch of a wind energy system is shown in Figure 10-6, and a solar energy system is shown in Figure 10-7. In both cases, energy from a wind farm or solar power plant is stored in compressed air energy storage

(CAES) systems. Off-peak energy is used to inject gas into the reservoir. A production well is used to withdraw gas to drive a turbine when wind or solar energy is not sufficient to meet demand.

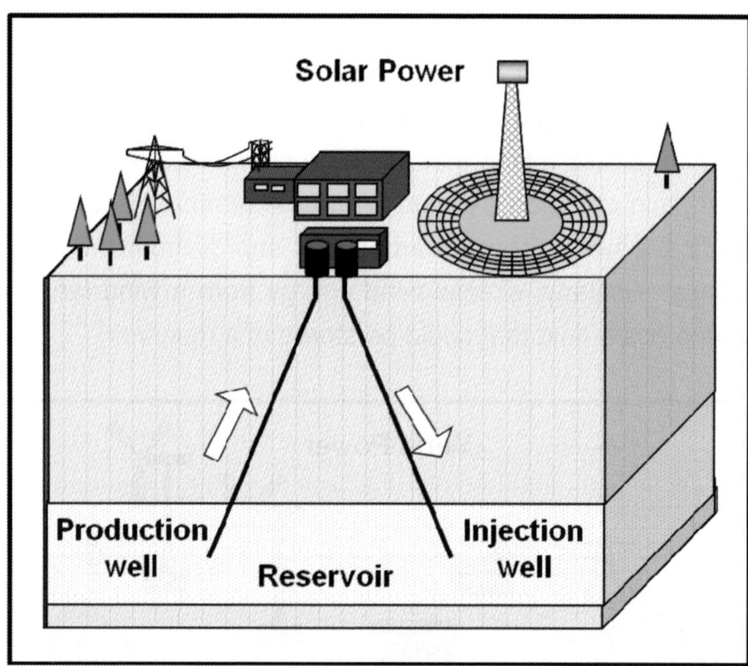

Figure 10-7. Sketch of CAES for a Solar Power Plant

10.4 A Hybrid Energy System for the Hydrogen Economy

A hybrid energy system is a system that combines two or more primary energy resources. For example, a wind farm supplemented by a natural gas fueled power plant is a hybrid energy system. Another example is the production and storage of hydrogen using an electricity-generating nuclear power plant.

Excess power capacity from sustainable energy systems such as nuclear plants, solar electric generating plants, and wind farms could be used to provide the energy needed to produce molecular hydrogen. The produced hydrogen would be suitable for a hydrogen-based economy, but the production of hydrogen requires a substantial amount of energy. Furthermore, hydrogen must be stored until it is needed. Figure 10-8 illuillustrates a hybrid system that could achieve the objectives of generating electricity, producing hydrogen, and storing hydrogen. The hybrid system includes a nuclear fission power plant and a hydrogen storage reservoir. Hydrogen is produced using off-peak power from the nuclear power plant. It is stored in the reservoir until it is needed.

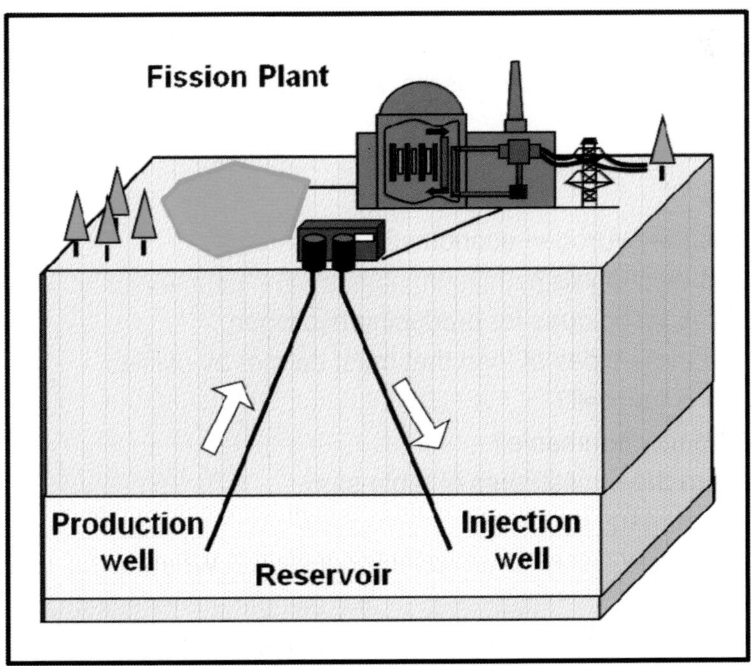

Figure 10-8. Sketch of a Nuclear Power Plant with Hydrogen Reservoir

10.5 Activities

True-False

Specify if each of the following statements is True (T) or False (F).

1. The *Hindenburg* exploded because the chemical coating on the outside of the zeppelin caused an explosion.

2. Molecular hydrogen can be produced from water by electrolysis.
3. The infrastructure exists for economically distributing and storing hydrogen.
4. Hydrogen is considered a source of energy.
5. The hydrogen atom is the first element in the Periodic Table.
6. Hydrogen is a common element found in sea water.
7. Thermal decomposition through steam reforming requires hydrocarbons and produces carbon dioxide.
8. In a fuel cell, hydrogen is fed to the cathode and oxygen is fed to the anode.
9. Hydrogen gas has a higher energy density than gasoline.
10. CAES facilities are considered viable large-scale energy storage options.

Questions

1. What is the hydrogen economy?
2. What is electrolysis?
3. List three techniques for producing hydrogen.
4. List the three types of load that must be met by utilities.
5. What is a fuel cell?
6. Is hydrogen flammable?
7. What are the constituents of table salt?
8. What is a metal hydride?
9. What are the benefits of superconductors? What is the main drawback?
10. Describe two methods of storing mechanical energy.

CHAPTER 11

ELECTRICITY GENERATION AND DISTRIBUTION

Power plants electrified the United States, and eventually the rest of the modern world, in only a century and a half. Although other types of energy are used around the world, electricity is the most versatile form for widespread distribution. The role of electric power plants is to generate electric current for distribution through a transmission grid. The historical developments that led to the modern power generation and distribution system are described below. We then discuss some changes that are being implemented to prepare electric distribution systems for the emerging energy mix.

11.1 HISTORICAL DEVELOPMENT OF ELECTRIC POWER

People first used muscle energy to gather food and build shelters. Muscle energy was used to grind grain with stones, chop wood with hand axes, and propel oar-powered ships. In many instances in history, conquered people became slaves and provided muscle energy for their conquerors.

Stones, axes, and oars are examples of tools that were developed to make muscle energy more effective. Water wheels and windmills replaced muscle power for grinding grain as long ago as 100 B.C. Wind and sails replaced muscle energy and oar-powered ships. Advances in the ability to use heat for power generation provided technological support for the industrial revolution.

11.1.1 Steam Generated Power

Early power stations were driven by wind and flowing water, and were built where wind and flowing water were available. Steam generated power allowed power plants to move, but had to await the development of furnaces. Furnaces used heat to smelt ore. Ore is rock that contains metals such as copper, tin, iron, and uranium. Heat from fire frees the metal atoms and allows them to be collected in a purified state. Copper and tin were the first metals smelted and could be combined to form bronze. Purified metals could be used separately or as alloys to build devices that could be operated at elevated temperatures. This technological advance was an important prerequisite for generating power using steam.

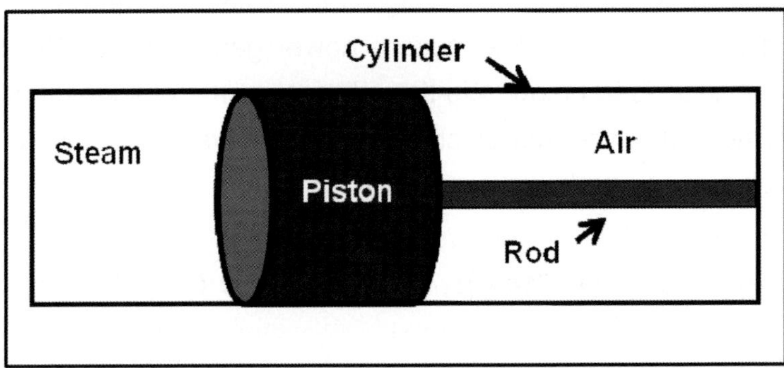

Figure 11-1. Schematic of a Simple Steam Engine

Scottish engineer James Watt improved the efficiency of the steam engine by introducing the use of a separate vessel for collecting and condensing the expelled steam. Watt's assistant, William Murdock, developed a gear system design in 1781 that enabled the steam engine to produce circular motion. The ability to produce circular motion made it possible for steam engines to provide the power needed to turn wheels. Steam engines could be placed on platforms attached to wheels to provide power for transportation. Thus was born the technology needed to develop steam-driven locomotives, paddle wheel boats, and ships with steam-driven propellers. The discovery that power plants did not have to be built near a particular fuel source meant that manufacturers had the freedom to build their manufacturing facilities in locations that optimized the success of their enterprise. If they chose, they could build near coal mines to minimize fuel costs, or near markets to minimize the cost of distributing their products.

Steam generated power was an environmentally dirty source of power. Burning biomass such as wood or a fossil fuel such as coal produced the heat needed to generate steam. Biomass and fossil fuels were also used in the home. Attempts to meet energy demand by burning primarily wood in 16th century Britain led to deforestation and the search for a new fuel source [Nef, 1977]. Fossil fuel in the form of coal became the fuel of choice in Britain and other industrialized nations. Coal gas, which is primarily methane, was burned in 19th century homes.

The demand for energy had grown considerably by the 19th century. Energy for cooking food and heating and lighting homes was provided by burning wood, oil, or candles. The oil was obtained from such sources as surface seepages or whale blubber. Steam generated power plants were providing power to consumers in the immediate vicinity of the power plant, but unable to serve potential customers that were not nearby. A source of power was needed that could be transmitted to distant consumers.

By 1882, Thomas Edison was operating a power plant in New York City. Edison's plant generated direct current electricity at a voltage of 110 volts. Nations around the world soon adopted the use of electricity. By 1889, a megawatt electric power station was operating in London, England. Industry began to switch from generating its own power to buying power from a power generating company. A fundamental inefficiency was present in Edison's approach to electric power generation, however. The inefficiency was not removed until the Battle of Currents was fought and won.

11.1.2 The Battle of Currents

The origin of power generation and distribution is a story of the Battle of Currents, a battle between two titans of business: Thomas Edison and George Westinghouse. The motivation for their confrontation can be reduced to a single, fundamental issue: how to electrify America.

Edison invented the first practical incandescent lamp and was a proponent of electrical power distribution by direct electric current. He displayed his direct current technology at New York City's Pearl Street Station in 1882. One major problem with direct current is that it cannot be transmitted very far without significant degradation.

Unlike Edison, Westinghouse was a proponent of alternating electric current because it could be transmitted over much greater distances than direct electric current. Alternating current could be generated at low voltages, transformed to high voltages for transmission through power lines, and then reduced to lower voltages for delivery to the consumer. Nikolai Tesla (1857-1943), a Serbian-American scientist and inventor who was known for his work with magnetism, worked with Westinghouse to develop alternating current technology. Westinghouse displayed his technology at the 1893 Chicago World's Fair. It was the first time one of the world's great events was illuminated at night and it showcased the potential of alternating current electricity.

The first large-scale power plant was built at Niagara Falls near Buffalo, New York in the 1890's. The power plant at Niagara Falls began transmitting power to Buffalo, less than 30 kilometers (20 miles) away, in 1896. The transmission technology used alternating current technology. The superiority of alternating current technology gave Westinghouse a victory in the battle of currents and Westinghouse became the father of the modern power industry. Westinghouse's success was not based on better business acumen, but on the selection of better technology. The physical principles that led to the adoption of alternating current technology are discussed below.

A chronology of milestones in the development of electrical power is presented in Table 11-1 [after Brennan, et al., 1996, page 22; and Aubrecht, 1995, Chapter 6]. The milestones refer to the United States, which was the worldwide leader in the development of an electric power industry.

Table 11-1
Early Milestones in the History of the Electric Power Industry in the United States

Year	Event	Comment
1882	Pearl Street Station, New York	Edison launches the "age of electricity" with his DC power station
1893	Chicago World's Fair	Westinghouse displays AC power to the world
1898	Fledgling electric power industry seeks monopoly rights as regulated utilities	Chicago Edison's Samuel Insull leads industry to choose regulation over "debilitating competition"
1907	States begin to regulate utilities	Wisconsin and New York are first to pass legislation
1920	Federal government begins to regulate utilities	Federal Power Commission formed

11.1.3 Growth of the Electric Power Industry

The power industry started out as a set of independently owned power companies. Because of the large amounts of money needed to build an efficient and comprehensive electric power infrastructure, the growth of the power industry required the consolidation of smaller power companies into a set of fewer but larger power companies. The larger, regulated power generating companies became public utilities and could afford to build regional electric power transmission grids. The ability to function more effectively at larger scales is an example of an economy of scale. Utility companies were able to generate and distribute more power at lower cost by building larger power plants and transmission grids.

Power plants operated by utilities should be able to meet three types of loads: base load, intermediate or cycling load, and peak load. Base load and peak load are minimum baseline and maximum demand respectively in a 24-hour period. Intermediate load is the demand that is required for several hours each day.

Many power plants are built near major population centers to reduce distribution costs. Electric power for small towns and rural communities was an expensive extension of the power transmission system that required special support. The federal government of the United States provided this support when it created the Tennessee Valley Authority (TVA) and Rural Electric Associations (REA).

Point to Ponder: Is society willing to adopt new technologies?

The history of electric power development shows that society is willing to adopt new technologies. We have seen this recently in the adoption of the world wide web, or internet, and cell phones. When people understand the benefits of new technology and an appropriate infrastructure is created, people will adopt the technology. It can be inferred, then, that society will accept changes in its current energy system when people understand the benefits and repercussions of any changes.

11.2 Electric Power Generation

The first commercial-scale electric power plants were hydroelectric plants. The primary energy source of a power plant is the energy used to operate an electricity-generating power plant. For example, flowing water is the primary energy source for a hydroelectric plant. Today, most electricity is generated by one of the following primary energy sources: coal, natural gas, oil, water, or nuclear. Table 11-2 presents the consumption of primary energy in 2006 as a percentage of total primary energy consumption in the world for a selection of primary energy types. The statistics give us an idea of the relative importance of different primary energy sources. Fossil fuels are the dominant primary energy source at the beginning of the 21st century. Electric energy, however, is the most versatile form of energy for running 21st century society. Much of the primary energy is consumed in the generation of electric energy.

Table 11-2
Primary World Energy Consumption in 2006 by Energy Type [US EIA website, 2009]

Primary Energy Type	Total World Energy Consumption
Oil	36.3 %
Natural Gas	22.9 %
Coal	27.0 %
Hydroelectric	6.3 %
Nuclear	5.9 %
Geothermal, Solar, Wind and Wood	1.8 %

11.2.1 Hydroelectric Power Generation

An example of electric power generation is hydroelectric power generation. Many of the first commercial electric power plants relied on flowing water as their primary energy source. People have known for some time that falling water could be used to generate electric power. A schematic of a hydroelectric power plant is presented in Figure 11-2. Hydroelectric power generation is discussed in detail in Chapter 8.

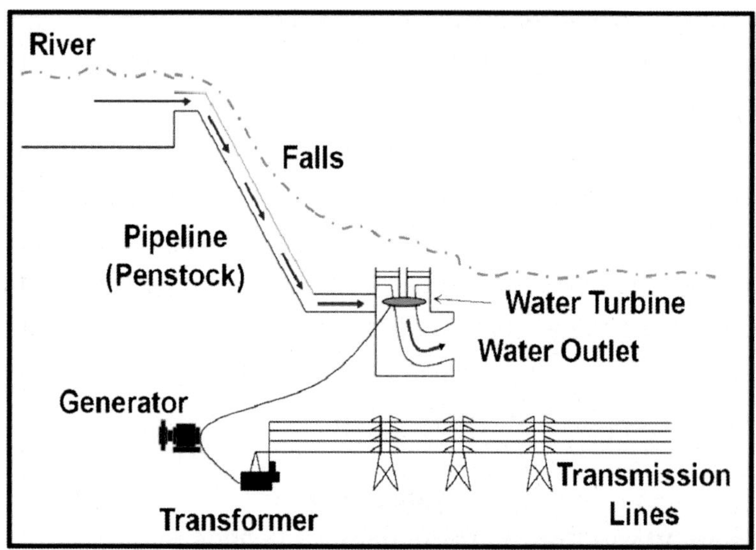

Figure 11-2. Principles of Hydroelectric Power Generation

Water flows from an upper elevation to a lower elevation through a pipeline called a penstock. The water current turns a turbine that is connected to a generator. The turbine is called the prime mover and rotates the generator shaft. The mechanical energy of flowing water is transformed into the kinetic energy of rotation of the turbine.

An alternating current generator converts mechanical energy to electrical energy using the following principle: an electric current is induced in a loop of wire that is allowed to rotate inside a constant magnetic field. If flowing water drives a turbine attached to a coil of wire inside a magnet, the coil of wire will rotate inside the magnetic field and an electric current will be induced in the coil of wire. Figure 11-3 shows a row of electric power generators at Hoover Dam, Nevada.

An electromotive force, or electric potential, is associated with the electric current. The electromotive force can be used to transmit electrical power through transmission lines. To do this efficiently, a transformer must be included in the system.

Figure 11-3. Electric Power Generators, Hoover Dam, Nevada (Fanchi, 2002)

11.3 TRANSFORMERS

Transformers perform the function of converting, or transforming, voltages. Voltage is the driving force for moving electrically charged particles. Current quantifies how many charged particles are moving. A transformer is a device that can convert a small alternating current voltage to a larger alternating current voltage, or vice versa. For example, it is desirable to work with relatively small voltages at the power plant and provide a range of voltages to the consumer. In between, in the transmission lines, a large voltage is required to minimize resistive heating losses in the line. A transformer that increases the voltage is referred to as a step-up transformer and is said to step-up the voltage; a transformer that decreases the voltage is referred to as a step-down transformer and is said to step-down the voltage. Transformer T_1 in Figure 11-4 is a step-up transformer from the low voltage (L.V.) at the power station to a higher voltage (H.V.) in the transmission line. Transformer T_2 in the figure is a step-down transformer that converts the relatively high voltage in the transmission line to a lower voltage that is suitable for the consumer. The low voltages shown at opposite ends of the transmission line do not have to be the same. The actual voltages used in the transmission line depend on the properties of the transformer.

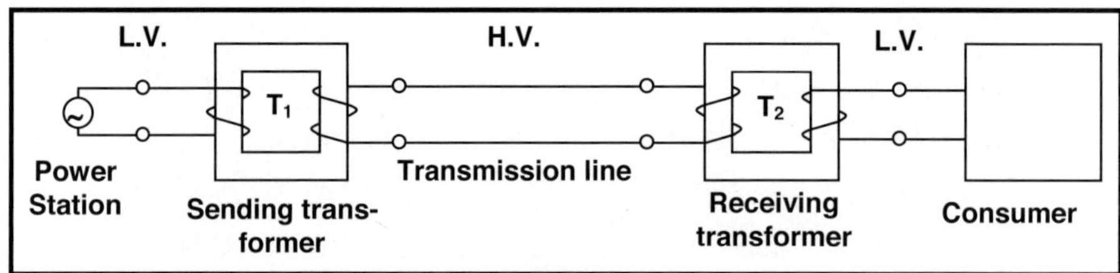

Figure 11-4. Power Transmission

Figure 11-5 shows that typical transmission line voltages can range from under 100,000 volts (100 kV) to over 750,000 volts (750 kV) [US-Canada Blackout Report, 2010, page 5]. The figure shows voltage step-up transformers between the generating station and transmission lines, and voltage step-down transformers between transmission lines and customers.

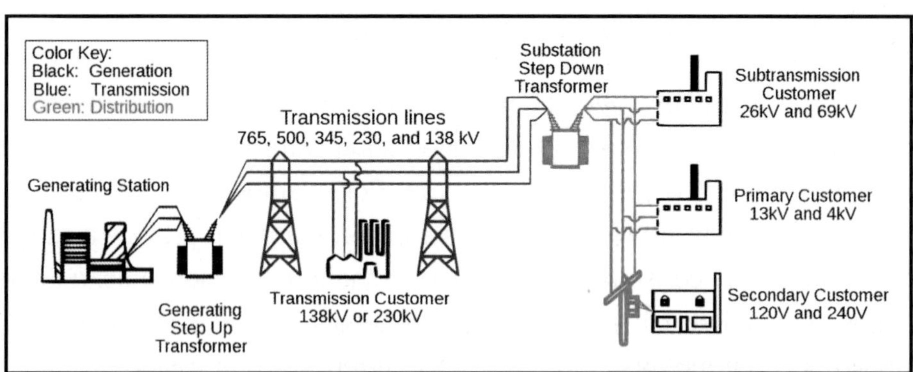

Figure 11-5. Schematic of a Power Transmission System

Advances in high voltage direct current (HVDC) technology have made it possible to transmit electrical power on HVDC power lines. Direct current can be transmitted at high voltages over great distances with relatively small power loss and cost compared to alternating current power transmission. In general, power is proportional to current, and power loss in the form of heat is proportional to the square of current. Thus, if we reduce current by a factor of two, we reduce power loss by a factor of four. For a given amount of power, an increase in voltage lets us reduce both the current and power loss associated with its transmission.

HVDC power transmission seems to be a desirable transmission technology, yet HVDC power transmission is not widely used. The problem is that many applications cannot operate using HVDC power. The transmission voltage needs to be reduced to be compatible with end-use applications such as lighting and motors. Modern semiconductor technology can be used to transform HVDC, but the existing infrastructure is designed with voltage transformers that work with alternating current electricity rather than direct current electricity. Converting the existing electric power distribution system to HVDC for all applications does not make economic sense. Consequently, HVDC power transmission has been limited to select applications, such as connecting electricity grids described below.

11.4 Electric Power Distribution

Transmission lines are used to distribute electric power. As a rule, society would like to minimize power loss due to heating to maximize the amount of primary energy reaching the consumer from power plants. We can reduce power loss by reducing the current or by decreasing the distance of transmission. In most cases, it is not a viable option to decrease the transmission distance. It is possible, however. For example, you could choose to build a manufacturing facility near a power station to minimize the cost of transmission of power. One consequence of that decision is that the manufacturer may incur an increase in the cost of transporting goods to market.

A more viable option for reducing power loss is to reduce the current that must be transmitted through transmission lines. Power loss increases faster with an increase in current than with an increase in transmission distance. This explains why Edison's direct current concept was not as attractive as Westinghouse's alternating current concept. The transmission of direct current incurs resistive power losses based on the direct current produced at the power plant. The purpose of transformers is to reduce the current in the transmission line, which is possible with alternating current.

11.4.1 Transmission and Distribution System

Electric power generating stations are connected to loads (consumers) by means of a transmission system that consists of transmission lines and substations. The substations are nodes in the transmission grid that route electric power to loads at appropriate voltages. Figure 11-6 shows the basic elements of an electric power transmission system. Typical transmission voltages in the United States range from 69 kilovolts to 765 kilovolts, and the alternating current frequency is 60 hertz. The infrastructure for providing electric power to the loads from the substations is the distribution system.

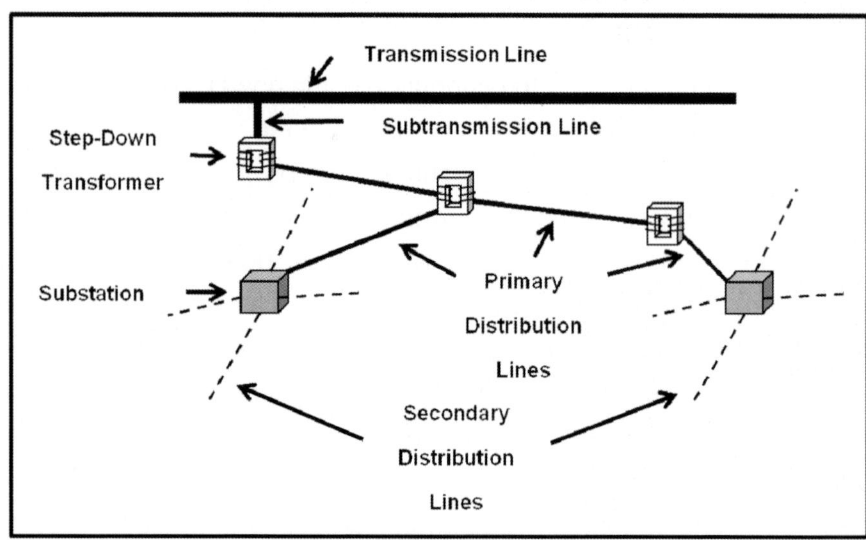

Figure 11-6. Power Transmission System

Electrical power can be transmitted over great distances using complex transmission systems. A failure in power transmission can affect large geographic areas and many people, as illustrated in Figure 11-7. The satellite images shown in Figure 11-7 are courtesy of the United States National Oceanic and Atmospheric Administration [US NOAA website, 2010]. They show the North American power blackout in August 2003. Power transmission failures can result in the loss of electrical power for millions of people.

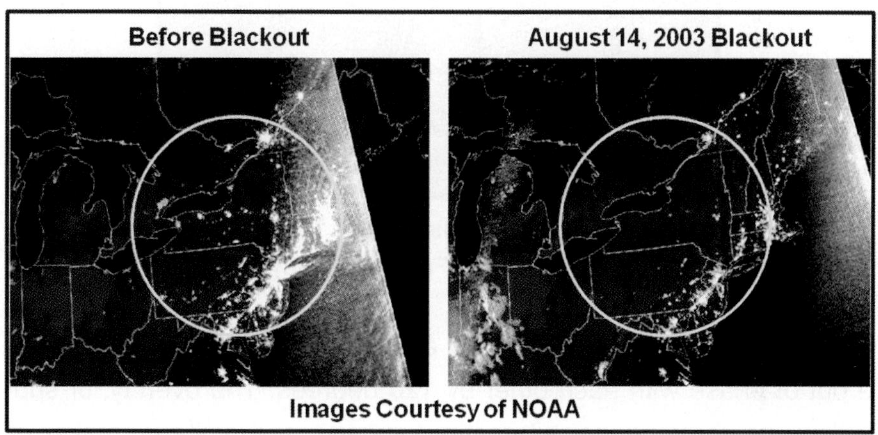

Figure 11-7. North American Power Disruption in August 2003

Power losses in transmission lines limit the distance that electrical power can be transmitted. An option for the future is to use superconductors as transmission lines. At low temperatures, resistance to electrical current vanishes in superconducting materials. Electrical power losses are minimized when resistance to an electrical current vanishes. Superconductors are not yet feasible for widespread use in power transmission because they require costly refrigeration.

Electric power is cost-effectively transmitted and distributed by operating a three-phase system. Transmission lines for three-phase electricity are shown in Figure 11-8.

Figure 11-8. Transmission of Three-Phase Electricity (Fanchi, 2003)

Three-phase electricity refers to alternating current and voltages that are out of phase with each other by 120 degrees. The overlay, or superposition, of three-phase alternating current and voltage makes the alternating current and voltage more uniform. A three-phase alternating current generator is used to provide three-phase electricity.

Three-phase electricity can be distributed to three one-phase loads using three separate conductors. The conductors are designed with small areas to minimize the amount of conducting material, such as copper, that must be bought and put into position. Three-phase transmission lines are designed to operate at high voltages and low currents. High voltage transformers are used to step-down the voltage for use by consumers. A typical pole-mounted transformer is shown in Figure 11-9.

A three-phase electrical transformer can convert three-phase electrical power at a distribution voltage of 13.8 kilovolts to single-phase voltages that are suitable for consumers. Residential consumers in the United States typically use single-phase voltages of 120 volts for small appliances and 240 volts for larger appliances such as clothes dryers.

Large, industrial consumers can use three-phase voltages on the order of 2160 volts or higher.

Figure 11-9. Pole-Mounted Transformer (Fanchi, 2003)

11.4.2 Household Circuits

Households are among the most common consumers of electricity in the modern world. Electricity is delivered as alternating current to a typical house in the United States using either a two-wire line or a three-wire line. The potential difference, or root mean square voltage, between the two wires in the two-wire line in the United States is 120 volts, and is 240 volts in many parts of Europe. One of the wires in the two-wire line is connected to a ground at the transformer and the other wire is the "live" wire. The three-wire line has a neutral line, a line at +120 volts, and a line at –120 volts.

A meter is connected in series with the power lines to measure the amount of electricity consumed by the household. Figure 11-10 shows series and parallel connections for three electrical devices. The series connection is designed so that the total current flows through each electrical device. The amount of current flowing through each electrical device in the parallel connection depends on the resistance of each device.

In addition to the meter, a circuit breaker is connected in series with the power lines to provide a safety buffer between the house and the power line. A fuse may be used instead of a circuit breaker in older homes. The fuse contains a metal alloy link, such as a lead and tin alloy, with a low melting temperature. If the alloy gets too hot because of resistive heating, the link will melt and break the circuit. More modern circuit breakers are electromagnetic devices that use a bimetallic strip. If the strip gets too hot, the bimetallic strip will curl based on the coefficients of thermal expansion of the two metals that make up the bimetallic strip. The change in shape of the strip when it curls breaks the circuit.

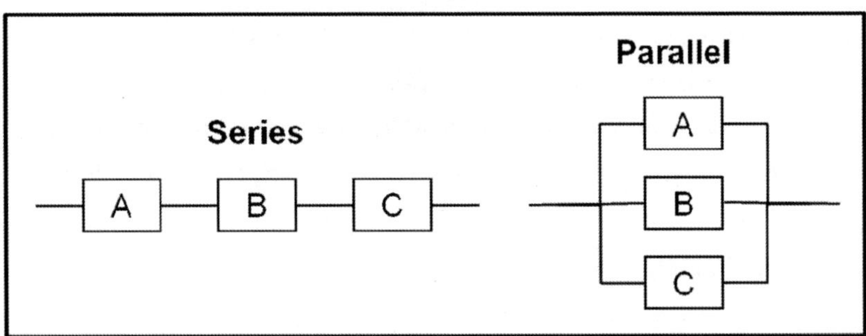

Figure 11-10. Schematic of Series and Parallel Connections

The power line and circuit breaker are designed to handle the current in the circuit, which can be as high as 30 amps. Many household applications require 15 amps of current. Lamps and appliances such as microwaves and toasters operate at 120 volts, while electric ranges and clothes dryers operate at 240 volts. Each circuit in the house has a circuit breaker or fuse to accommodate different loads.

The circuit in the house is connected in parallel to the power lines. The parallel connection makes it possible to turn on and shut off an electrical device without interfering with the operation of other electrical devices. If a series connection were used, the circuit would be broken whenever one of the electrical devices was turned off. That is why circuit breakers are connected in series between the household circuit and the power lines; the circuit breaker is designed to disconnect all electrical devices in the house from the power lines in the event of an overload, such as a power surge. An open circuit occurs when the continuity of the circuit is broken. A short circuit occurs when a low resistance pathway is created for the current.

Electricity can be harmful if a person touches a live wire while in contact with a ground. Electric shocks can cause burns that may be fatal and can disrupt the routine functioning of vital organs such as the heart. The extent of biological damage depends on the duration of contact with the current and the magnitude of the current. A current in excess of 100 milliamps (0.1 amp) can be fatal if it passes through a body for a few seconds.

Electrical devices and power lines should be handled with care. Three-pronged power cords for 120-volt outlets provide two prongs that are grounded and one prong that is connected to the live wire. The grounded prongs are provided for additional safety in electrical devices designed to use the three-pronged cords. One of the ground wires is connected to the casing of the appliance and provides a low resistance pathway for the current if the live wire is short-circuited.

11.4.3 Distributed Generation

Practical considerations limit the size of power plants. Most large-scale power plants have a maximum capacity of approximately 1000 megawatts. The size of the power plant is limited by the size of its components, by environmental concerns, and by energy source. For example, the area occupied by a power plant, called the plant footprint, can have an impact on land use. Conventional power plants that burn fossil fuels such as coal or natural gas can produce on the order of 1000 megawatts of power. Power plants that depend on nuclear reactors also produce on the order of 1000 megawatts of power. By contrast, power plants that rely on solar energy presently can produce on the order of 10 megawatts of power. Power from collections of wind turbines in wind farms can vary from one megawatt to hundreds of megawatts. If we continue to rely on nuclear or fossil fuels, we need to work with power plants that generate on the order of 1000 megawatts of power. If we switch to power plants that depend on solar energy or wind energy, the power generating capacity of each plant is less than 1000 megawatts of power and we must generate and transmit power from more plants to provide for existing and future power needs.

In some areas, public pressure is growing to have more power plants with less power generating capacity and more widespread distribution. The federal government of the United States passed a law in 1978 called the Public Utilities Regulatory Policies Act (PURPA) that allows non-utilities to generate up to 80 megawatts of power and requires utilities to purchase this power. PURPA was the first law passed in decades to relax the monopoly on power generation held by utilities and reintroduce competition in the power generating sector of the United States economy.

Historically, distributed generation was the first power generation technology. The distributed generation of energy is the generation of energy where it is needed and on a scale that is suitable for the consumer. Examples of distributed generation include a campfire, a wood stove, a candle, a battery-powered watch, and a car. Each of these examples generates its own power for its specific application. Distributed generation was replaced by a system that relies on large-scale power generating plants and extensive transmission capability. The modern electricity grid transmits electrical power to distant locations.

Some people believe that the future of energy depends on a renaissance in distributed generation. In this view, a few large-scale power plants in the centralized system will be replaced by many smaller-scale power-generating technologies. A. Borbely and J.F. Kreider define distributed generation as "power generation technologies below 1 megawatts of electrical output that can be sited at or near the load they serve." [Borbely and Kreider, 2001, page 2] This definition does not include small-scale power-generating technologies whose ideal locations depend on the locations of their energy source. For example, hydropower and wind-powered generators are not considered distributed generation technologies according to Borbely and Kreider's definition because hydropower and wind-powered generators depend on the availability of wind and water sources that are suitable for hydroelectric plants and wind farms. Consequently, hydropower and wind-powered generators are located near their energy sources. The locations are often not near the power consumer.

11.5 Developments in Electric Grid Design

We have seen that power from wind and sunshine is not always produced when it is needed. The sun does not shine on a particular spot on the surface of the earth all day long, but the sun does shine all the time. We can use sunlight 24 hours per day if we can collect sunlight at different locations around the globe. Similarly, the wind does not usually blow all the time at a particular location, but wind is always blowing somewhere. Distributed generation can be used to harvest both wind energy and solar energy. Improved power distribution and transmission systems could be used to produce energy from wind and sunshine in different parts of the world and transmit it to places where it is needed. A regional or global energy distribution system would have to be developed and maintained to achieve this capability. Advances in grid design are discussed here.

11.5.1 North American Grid

The contiguous United States electrical grid is part of the North American grid. The North American grid consists of three alternating current subgrids connected by high voltage direct current (HVDC) lines. The three subgrids include the Eastern Interconnect region, the Western Interconnect region, and the Texas Interconnect region that includes most of Texas. The interconnect regions are shown in Figure 11-11 [DoE Power

Grid, 2010]. The Texas and Western Interconnect regions are linked with the Mexican grid, and the Eastern and Western Interconnect regions are connected with the Canadian grid.

The Eastern, Western and Texas Interconnect regions are subdivided into 10 North American Electric Reliability Council (NERC) regions. The NERC regions are ECAR (East Central Area Reliability Coordination Agreement), ERCOT (Electric Reliability Council of Texas), FRCC (Florida Reliability Coordinating Council), MAAC (Mid-Atlantic Area Council), MAIN (Mid-America Interconnected Network), MAPP (Mid-Continent Area Power Pool), NPCC (Northeast Power Coordinating Council), SERC (Southeastern Electric Reliability Council), SPP (Southwest Power Pool), and the WSCC (Western Systems Coordinating Council). The NERC regions are shown in Figure 11-11. Alaska and Hawaii are separate systems. The Hawaiian Islands, in particular, use minigrids to provide power to each island.

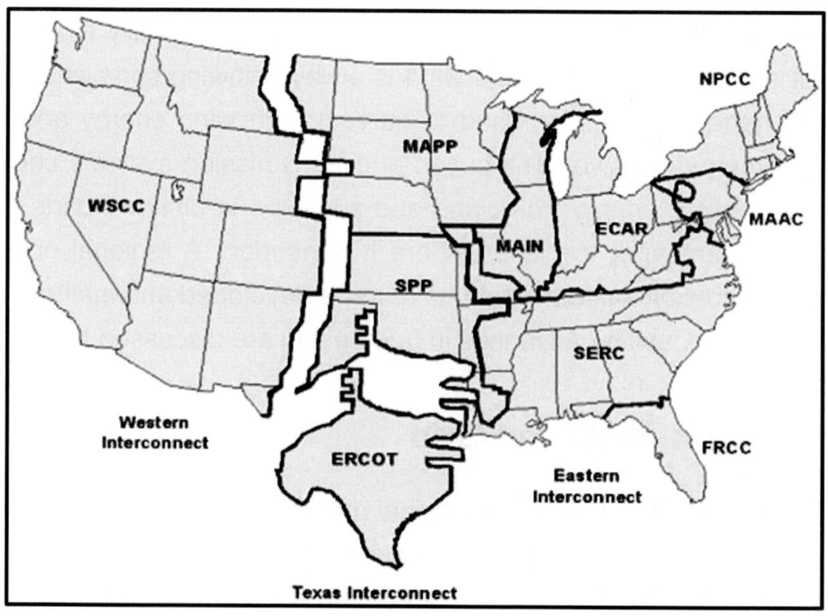

Figure 11-11. North American Electricity Grid

Electric grids connect consumers to electrical power. In addition to connecting power stations with consumers, transmission lines provide a level of redundancy so that demand can be met by complementary power sources. This does not mean that system failure cannot occur, as we saw with the 2003 blackout described in Section 11.4.1.

11.5.2 Smart Grid

The United States Federal Energy Regulatory Commission (FERC) has authority over the interstate transmission of electric power. It is charged with modernizing the United States electric system under the auspices of the 2007 Energy Independence and Security Act [US FERC website, 2010]. FERC is developing a Smart Grid which is expected to apply digital technologies to the grid, and enable real-time coordination of information from generation supply resources, demand resources, and distributed energy resources.

Title XIII of the Energy Independence and Security Act of 2007 specifies characteristics of a Smart Grid [US FERC website, 2010]:

1. Increase use of digital information and controls technology to improve reliability, security, and efficiency of the electric grid.
2. Dynamically optimize grid operations and resources, with full cyber-security.
3. Deploy and integrate distributed resources and generation, including renewable resources.
4. Develop and incorporate demand response, demand-side resources, and energy efficiency resources.
5. Deploy "smart" technologies (real-time, automated, interactive technologies that optimize the physical operation of appliances and consumer devices) for metering, communications concerning grid operations and status, and distribution automation.
6. Integrate "smart" appliances and consumer devices.

7. Deploy and integrate advanced electricity storage and peak-shaving technologies, including plug-in electric and hybrid electric vehicles, and thermal storage air conditioning.
8. Provide consumers timely information and control options.
9. Develop standards for communication and interoperability of appliances and equipment connected to the electric grid, including the infrastructure serving the grid.
10. Identify and lower unreasonable or unnecessary barriers to adoption of smart grid technologies, practices, and services.

Point to Ponder: Should we be concerned that the Smart Grid will infringe on our freedom?

Some people are concerned that they will lose some of their freedom when the Smart Grid is fully implemented. The source of this concern is the FERC charge to provide "real-time coordination of demand resources." If demand resources include businesses and households, then FERC could regulate how businesses and households use energy for such applications as heating and cooling. The Smart Grid could ration power or limit your use of power at inconvenient times without special authorization. For example, if you need power for a special event such as a conference or party, you might have to obtain authorization from a utility or governmental agency. The need to obtain authorization can infringe on existing liberties and gives some people authority to allocate resources. This increases the possibility that corruption will occur. Items 8 and 10 of the Smart Grid characteristics listed above can be used to alleviate these concerns, but are unlikely to eliminate corruption.

11.5.3 European Super Grid

The Union for the Coordination of Transmission of Electricity (UCTE) began unifying the electricity grids of the European Union member states in the 1950s after World War II. The UCTE was formed in 1951 as the Union for the Coordination of Production and Transmission of Electricity

(UCPTE). In 1999, UCPTE became an association of transmission system operators (TSOs) called the UCTE when the EU deregulated electricity markets. Members of UCTE include most of continental Europe. Electricity grids in the North African states of Morocco, Algeria and Tunisia are synchronized with the UCTE grid through the Gibraltar AC link.

UCTE is now a member of the European Network of Transmission System Operators for Electricity (ENTSO-E). Membership in ENTSO-E includes more TSOs and more countries than the UCTE, including TSOs from Great Britain, which was not a member of UCTE. The mission of ENTSO-E has several purposes [ENTSO-E website, 2010]: pursue the co-operation of European TSOs on regional and pan-European levels; promote the interests of member TSOs; and have an active and important role in the European rule-setting process in compliance with EU legislation.

Plans exist for altering the EU grid to a Super Grid (or Mega Grid) that encompasses all EU member states. The Super Grid is advocated by DESERTEC Foundation, an initiative of the Club of Rome. The Club of Rome was founded in 1968 by Italian industrialist Aurelio Peccei and Scottish scientist Alexander King. It is a global not-for-profit organization with the mission of acting "as a global catalyst for change through the identification and analysis of the crucial problems facing humanity and the communication of such problems to the most important public and private decision makers as well as to the general public." [Club of Rome, 2010, see Organization]. The concept of a Super Grid is also supported by the European Wind Energy Association (EWEA) since this grid should help connect large planned North Sea wind farms to European consumers. The EWEA sponsored a 2010 conference on the development of an improved European electricity grid [http://www.ewea.org/grids2010/].

The Super Grid planned by DESERTEC is expected to transmit electrical power generated by offshore wind farms in the North Sea and electrical power generated by Spanish and North African solar power plants [Desertec Red Paper, 2010; Desertec White Paper, 2010]. The plan includes construction of desalination plants to provide potable water to North African and Middle Eastern nations. The Super Grid will connect the EU with Saharan Africa. Critics of the plan express concern that the project will make the EU dependent on energy from non-member states, and will be vulnerable to attack.

11.6 Activities

True-False

Specify if each of the following statements is True (T) or False (F).

1. George Westinghouse was a proponent of alternating current.
2. The load on a utility is the demand for electric power.
3. A generator can convert mechanical energy to electrical energy.
4. Thomas Edison proposed using alternating current to distribute electric power.
5. Steam engines were developed to convert thermal energy into mechanical energy.
6. Thomas Edison launched "the age of electricity" at the Pearl Street station in 1882 with his DC power station.
7. Electrical power can be transmitted over great distances without loss of power using the existing transmission grid.
8. The first commercial-scale electric power plants were hydroelectric plants.
9. A short circuit occurs when the continuity of the circuit is broken.
10. The European Super Grid will include wind power from the North Sea region and solar power from Spain and North Africa.

Questions

1. Suppose the minimum daily power provided by a utility is 100 MW and the maximum daily power is 50% greater than the minimum. Express the base load and peak load of the utility in MW.
2. Is electrical power transmitted over transmission lines using low voltage or high voltage? Explain your answer.
3. What does an alternating current transformer do?

4. Is the voltage used in household circuits in the United States the same as the voltage used in Europe?
5. What is distributed generation?
6. What is the efficiency of a power station that uses 120 MW power to generate 100 MW power?
7. How many phases used to transmit and distribute electric power in the United States?
8. What two systems are connected in series to power lines that enter households? What are their purposes?
9. What are the three alternating current subgrids called in the North American grid?
10. How might a "smart" grid interfere with personal and business freedom?

CHAPTER 12

ENERGY ECONOMICS

Energy may be the most important factor that will influence the shape of society in the 21st century. The cost and availability of energy significantly impacts our quality of life, the health of national economies, the relationships between nations, and the stability of our environment. What kind of energy do we want to use in our future? Will there be enough? What will be the consequences of our decisions? The selection of an energy source depends on such factors as availability, accessibility, environmental acceptability, capital cost, and ongoing operating expenses. We discuss how economic factors affect decision-making in this chapter. Many of the concepts presented here are discussed in standard economics textbooks, such as Varian [2009] and Dahl [2004].

12.1 Principles of Economics

Economic decisions are made on a microeconomic or macroeconomic scale. Microeconomics is the study of how individual businesses or households decide to allocate limited resources. Macroeconomics is the study of the behavior of economic units that are aggregated into community, national or global economies. Both the decision-making of individual economic units and the policy-making of governments depend on the supply of and demand for energy.

Prices of goods and services in a market economy are determined by supply and demand. For comparison, in a planned economy prices of goods and services are set by a central government. A market is a venue where goods and services can be exchanged for other goods and services or for currency. The rate of exchange, or the value of the goods or services being exchanged, is considered the price and can be quantified using a currency such as U.S. dollars.

When a country or other decision-making body considers which energy sources are best suited for them, it is important to not only examine the price today, but the potential price in the future. For most countries, the determining factor for deciding which energy sources to use is price. Factors such as environmental impact, political impact, and national security can be factored into price.

12.1.1 Supply and Demand

Supply and demand are expressed by the relationship between price and quantity of units. Figure 12-1 illustrates supply and demand curves. The plot of supply and demand is a snapshot in time. The demand curve slopes down from left to right because normal market behavior is to buy more of a product when the price of a unit is low and buy less when the price is high. The supply curve tends to slope up from left to right because normal market behavior is to provide less of something when the price is low and provide more when the price is high. The difference between price of a unit and the cost of a unit is profit.

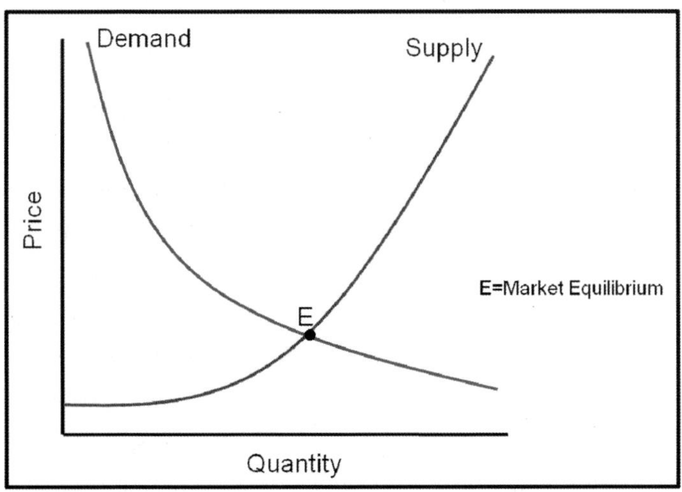

Figure 12-1. Supply and Demand Curves

The economic theory of supply and demand tells us how quantity of units is affected by price. Figure 12-2 shows how a change in quantity influences price with regard to demand. The base case labeled 12-2a relates price and a specified quantity of units. An increase in price in case 12-2b is accompanied by a decrease in quantity sold. On the other hand, a decrease in price in case 12-2c is accompanied by an increase in quantity sold. A similar discussion can be applied to supply. In that case, an increase in price is accompanied by an increase in quantity available for sale. A decrease in price is accompanied by a decrease in quantity available for sale.

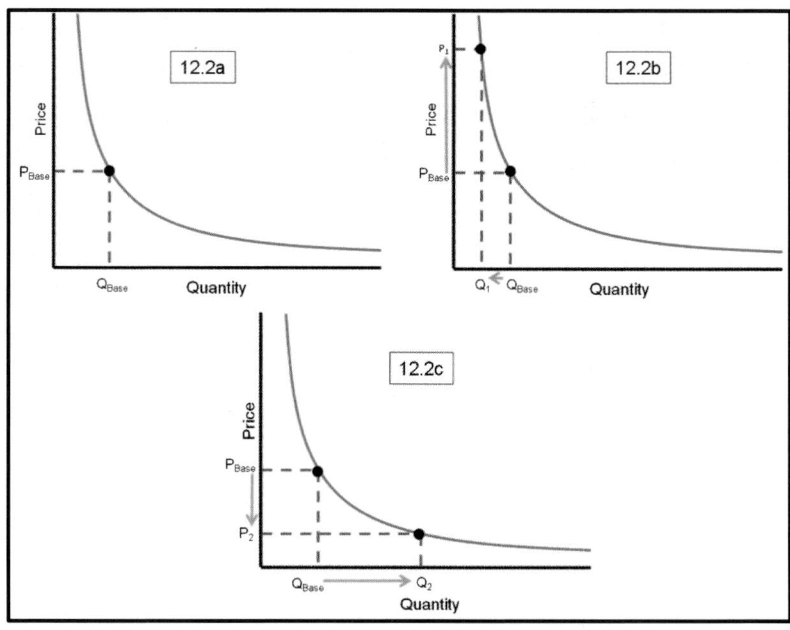

Figure 12-2. Demand and Price Relationship

The point on the curves where supply and demand intersect is called market equilibrium. This is the point that markets seek to achieve, where the amount produced (supplied) is equal to the amount consumed (demanded). Demand does not need to come entirely from private consumers and businesses. Governments can be consumers of goods and services as well, and this is included in the demand curve. Demand also does not necessarily imply that all of a good will be consumed. The demand for goods often varies with time. Consequently, the supply and demand curves for many goods and services change significantly as a function of time. For example, in the United States, the demand for heating oil increases during the winter, and the demand for gasoline increases during the summer vacation season.

12.1.2 Elasticity

The slope of the demand curve is a measure of the elasticity of demand. Elasticity of demand refers to the sensitivity of demand to changes in price and can be viewed as elastic or inelastic. Elastic demand and inelastic demand are compared in Figure 12-3. Demand is elastic when a price change is accompanied by a corresponding change in quantity sold. Demand is inelastic when a price change has little or no impact on demand.

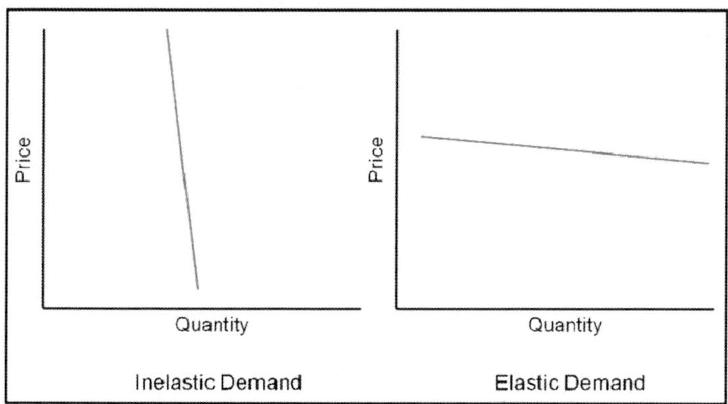

Figure 12-3. Elasticity of Demand

There are several possible reasons that demand for a product may be elastic. For example, if a product has a very close substitute or many potential substitutes, then an increase in price of that product will motivate consumers to switch to one of the other products. This is called product substitution. Another explanation for elastic demand is that a product may be a discretionary purchase, that is, the product has a limited number of specific uses that are not considered necessities. Thus an increase in price will cause a large drop in consumer interest or a willingness to delay the purchase.

Inelastic demand occurs when price can change significantly without having a significant impact on demand. For example, in recent years, the price of oil in the United States has seen its greatest percentage increase since the first oil crisis of the 1970s. From May 2003, when the average price of a barrel of crude oil was roughly US$23, to July 2008 when the average price per barrel was US$135, crude oil prices rose roughly 586%. While the price increased significantly, the demand for petroleum products in the United States did not significantly decline. United States petroleum use (demand) was roughly 600 million barrels in May 2003, and was 606 million barrels in July 2008, a 1% increase. With oil prices affecting most Americans at the gas pump, many changed their driving habits as much as they could, but the country still required vast quantities of oil to function. In this example, demand for oil in the price range from US$23 per barrel to US$135 per barrel was inelastic, that is, demand for oil was largely unaffected by increases in price.

The elasticity of a curve can vary depending on price. This was illustrated in Figure 12-2 above, where the slope of the demand curve was steepest at the top, meaning that demand at that quantity level is inelastic. As price drops, the demand became more elastic. This figure illustrates demand for oil in the United States. When the price is highest, demand decreases but there is a quantity of oil that will be demanded with minimal sensitivity to price. As the price gets lower, however, Americans tend to use oil in increasing amounts because there are very few substitutes for oil, particularly for transportation, and oil has many different uses.

The existence of substitutes is one of the largest determining factors for price elasticity. That is why each oil crisis over the past 40 years has been met with a call to decrease the United States' dependence on foreign oil. Figure 12-4 shows that the United States imports more than half of the oil it consumes. The cost of energy in the United States is sensitive to price setting decisions by foreign bodies. One of the first demonstrations of this was the 1973 oil crisis when OPEC enacted an oil embargo against the United States and caused an unprecedented increase in oil price globally. OPEC and non-OPEC oil exporting countries have a strategic advantage over oil importing countries.

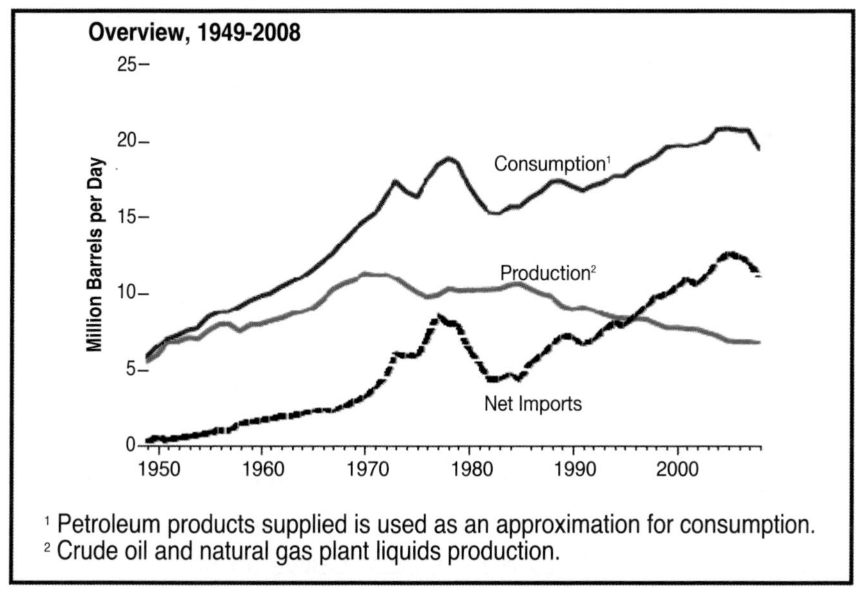

Figure 12-4. U.S. Petroleum Consumption, Production and Net Imports [eia.doe.gov, accessed 11/25/09]

According to the United States Energy Information Administration, the United States, Japan, China, Germany and South Korea are the top five net oil importers in the world. The economies of these countries are particularly vulnerable to price changes. It is no surprise that each of these countries is interested in developing commercially competitive alternative fuels. The United States and Germany are developing wind and solar power. Japan and South Korea are developing solar power. China is developing wind and solar power, and implementing hydropower.

12.1.3 CAPEX and OPEX

In the near term, such as the next two to four years, the cost of solar energy is going to be too high for countries to consider it a viable primary energy source. The cost of commercializing solar energy is increased by the significant cost of constructing power plants and related infrastructure to distribute the energy to consumers. These costs are known as capital expenditures, or CAPEX, because they are incurred at the beginning of the project but will create future benefits.

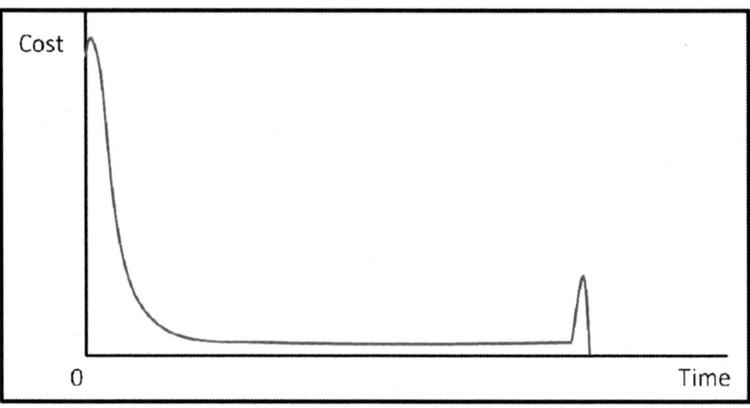

Figure 12-5. Cost Curve

While price has a large impact on supply and demand in the near term, price can change in the future. When evaluating energy sources like wind and solar, it is likely that the price will be significantly lower in the future. Once the production facility has been built, the cost of fuel is negligible and long term costs are operational expenses, or OPEX. OPEX includes fuel, personnel, and maintenance costs. Although some alternative power sources would be considered too costly today, over the life of the equipment, they may be cheaper than other energy options. Figure 12-5 illustrates a typical cost curve for a project. At the beginning, the cost is highest because the basic facility must be built. Cost then stabilizes at a lower level when the facility is completed and the only costs are OPEX. At the end of the life of the facility, the cost to decommission the facility and repair or mitigate any environmental damage leads to an increase in cost.

When making economic decisions regarding energy, it is important to consider substitute products such as alternative energy. Many countries are making decisions today to construct expensive alternative energy facilities such as wind farms, solar power plants and hydroelectric dams because they are concerned that "peak oil" is imminent or has already occurred. Theories about peak oil tend to vary widely. Some claim that peak oil has already been reached, while others argue that it may be decades before peak oil is reached. Access to oil is one factor that needs to be considered by a country that is deciding whether to implement policies advocating the development of alternative energy sources. Other key considerations include environmental impact, energy independence and security. Chapter 4 discusses peak oil in greater detail.

12.2 Costs and Benefits

Energy is not free. The financial cost of energy comes in two forms, direct and indirect. The societal cost of energy can be more difficult to quantify. Direct and indirect costs, and tangible and intangible costs and benefits are explained here.

12.2.1 Direct and Indirect Costs

Costs to energy producers come in two forms: direct and indirect costs. Direct costs are the costs directly associated with producing and transporting energy to consumers. This includes the cost of materials and labor to construct the facility, and operating costs. Fuel costs are direct costs that are significant costs for coal, oil, gas, and nuclear power plants, but are negligible for solar, wind, and hydroelectric plants. Many of these power plants use large volumes of water that must be purchased as another direct cost.

A significant up-front direct cost of new alternative energy projects is the cost of financing for these projects. When a company is seeking to build a power facility using a relatively new technology, that technology is considered to be riskier than more established technologies. Thus, the company typically pays a higher interest rate on any money borrowed to build the facility because of the lender's perception that the risk of failure to function or meet expectations is greater.

Direct costs are costs incurred for activities or services that provide benefits to specific projects. Indirect costs are costs that benefit multiple projects. Overhead is an indirect cost. The term "overhead" typically refers to expenses required for a business to function, but do not directly generate profits. Overhead includes expenditures for managing facilities, paying personnel, and paying rent and utilities. The following example illustrates the concept of overhead.

Suppose an energy company owns a wind farm, a coal-fired power plant, and a natural gas power plant. Such a company would typically employ a number of people to account for the costs and revenues related to the wind farm and each of the power plants. The expense for this accounting work can be proportionally divided among the three energy systems, but it is not feasible to determine exactly how much accounting time is dedicated to each system. The cost of accounting is an indirect cost. Office buildings that contain office support services such as accounting and human resources, as well as front office services such as customer service, are also considered indirect costs.

Indirect costs can include permitting and other governmental costs that a company must pay. When a company chooses to build a new power plant, they must receive authorization from local and national government agencies. The cost of the permits and the costs of preparing any related paperwork are considered indirect costs. An indirect cost for companies operating in some countries is the cost of influencing officials who have the power to grant permits or mineral extraction rights.

Another type of cost is opportunity cost. Opportunity cost is the concept that the selection of one opportunity that requires the use of scarce resources means another opportunity is not selected. The value of the opportunity that was not selected is the opportunity cost. For example, suppose company A has enough money to build several wind turbines or they can use the money to drill for natural gas, but they cannot do both. If the company chooses to drill for gas, the wind turbines that go un-built and the related revenues from those turbines are the opportunity cost to the company for choosing the natural gas project.

12.2.2 Tangible and Intangible Costs and Benefits

Economic decisions should consider tangible financial costs as well as intangible costs. Intangible costs are the costs that tend to be difficult to quantify and can be financial or societal. For example, the environmental impact of an energy source can be measured with both tangible and intangible costs. Tangible costs would include the cost of installing environmental protection measures such as scrubbers in coal plants, radiation barriers in nuclear power plants, or sound dampening walls around drilling rigs in urban areas. Tangible environmental costs also include the projected costs of cleaning up a site, whether that clean-up is the result of a chemical spill, or the cost of reclaiming a drill site or power plant site after it is decommissioned.

Intangible costs include factors such as carbon emissions and their impact on the atmosphere. Energy producers can estimate the financial cost to decrease emissions, but the damage to the environment can be difficult, if not impossible, to quantify. Society as a whole must consider the societal cost of using an energy source that produces these emissions. Another intangible cost can be the aesthetics related to an energy source.

Oil refineries, coal power plants, and even wind farms are considered unattractive by some people. Aesthetics is an intangible cost associated with those energy sources. An intangible cost that governments and energy companies must pay close attention to is the public's or consumer's feelings about an energy source. For example, fossil fuels have received significant negative publicity because of the growing concern about greenhouse gas emissions. An unquantifiable cost to countries continuing to use fossil fuels is the negative publicity that is associated with these fuels domestically and internationally.

Energy sources have tangible and intangible benefits that must be considered as well. Tangible benefits include something as simple as price if one energy source is cheaper than another. Tangible benefits also include effectiveness. Gasoline, for example, is effective as a transportation fuel. This is a tangible benefit for oil relative to other sources of energy. Intangible benefits include the environmental impact of the energy source. Unlike fossil fuels, which produce undesirable carbon emissions, many alternative energy sources have a good public image because they do not produce carbon emissions. The lack of carbon emissions is a tangible benefit for wind farms, solar power plants, and hydroelectric plants because it eliminates the cost of mitigating greenhouse gas emissions, and the associated image of being a "clean" energy source is an intangible benefit.

12.3 Economies of Scale

The cost of wind and solar energy is relatively high compared to fossil energy. This is due in part to the higher cost of production using technology that is still developing. As the demand for a product increases, producers can expand production and increase their profit by improving the technology and reducing the average cost of the product. The reduction of the average cost of a product as the scale of production expands is referred to as an economy of scale. Economies of scale are achieved if the average cost of a product decreases when a firm expands the production of the product. The history of the modern wind turbine industry illustrates the concept of economies of scale.

Interest in renewable energy, including wind turbines, was stimulated in the United States and Europe when people realized that oil supply could be used as a political weapon following the 1973 oil crisis. The most important result of that oil crisis in the context here is that the price of oil increased to the point that leaders in the United States and Europe began to understand that dependence on a foreign energy supplier carried a significant cost. Domestic access to energy resources has considerable political value. This realization has been the impetus for renewable energy research since the 1973 oil crisis.

The cost of renewable energy was prohibitively high and interest in renewable energy waned when the price of oil stabilized after the end of the crisis in early 1974. Several oil supply disruptions have occurred since 1973, and the demand for renewable energy has continued to grow. Wind turbine technology in particular developed quickly and is now economically competitive. Demand for wind turbines has increased, wind turbine technology has improved, and production capacity has expanded. Today, the number of commercial wind turbines is increasing and the cost for manufacturing and installing wind turbines has decreased. This is an example of economies of scale. Wind energy has become economically competitive with fossil energy and is a viable technology for energy providers to consider adding to their energy portfolio.

Recognition that energy from oil can be replaced by energy from another energy source can be expressed in terms of the concept of fungibility. We use a specific example to clarify the concept. In particular, electrical energy can be provided using several different energy sources. If the price of one energy source gets too expensive, a cheaper source can be substituted for the more expensive source. This is product substitution, and is possible because electrical energy is fungible, that is, electrical energy produced by one type of energy source, such as burning oil or coal, is interchangeable with electrical energy produced by another type of energy, such as harvesting wind or solar energy.

A unit of a product is fungible if individual units of the product are interchangeable. Thus a kilowatt-hour of electricity from the burning of oil is interchangeable with a kilowatt-hour of electricity from wind harvested by a wind turbine. Electricity is a fungible commodity. It can be produced from any of a large number of energy sources and offered to the consumer at a price that optimizes the cost to the producer. Consumers can use electricity without knowing how their electricity was produced or where it came from. Other fungible energy products include a barrel of oil, a cubic foot of natural gas, a pound of coal, and a pound of uranium-235.

12.4 MANAGEMENT DECISIONS IN THE ENERGY SECTOR

Decision makers at energy companies and within governments must consider many economic factors when selecting energy projects and formulating policy. Energy projects range from drilling a well to building expensive, offshore platforms or solar power plants, or even venturing into a new energy type entirely. Some of the concepts involved in making management decisions are presented here.

12.4.1 Cash Flow and Economic Indicators

An economic analysis of competing investment options usually requires preparing cash flow predictions. The cash flow of an investment option is the net cash generated by the investment option or expended on the investment option as a function of time (see Figure 12-6). The value of money can change with time. The time value of money is included in the economic analyses by applying a discount rate to adjust the value of money in future years to the value of money during a base year. The resulting cash flow is called the discounted cash flow. The net present value (NPV) of the discounted cash flow is the value of the cash flow at a specified discount rate.

Net present value is the difference between revenue and expenses. Net present value, revenue, and expenses depend on the time value of money. We can account for the time value of money by introducing a discount rate in the calculation. Revenue can be calculated from the price per unit produced times the quantity produced. The quantity produced can be volume of oil or gas, kilowatt-hours of electricity, or any other appropriate measure of resource production. Expenses include capital expenditures, operating expenditures, and taxes.

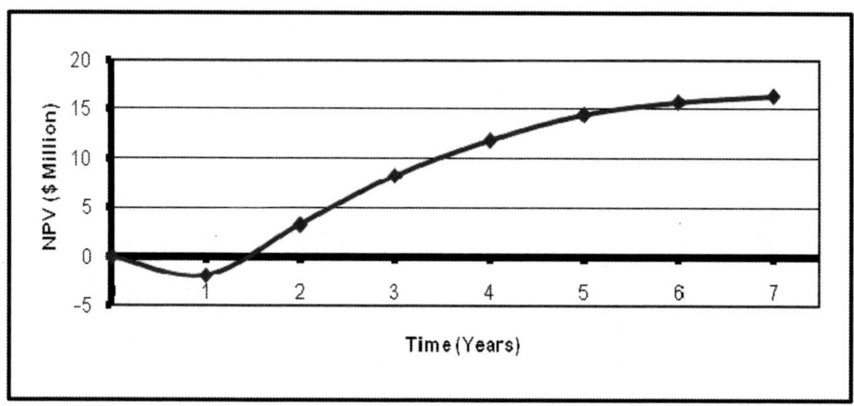

Figure 12-6: Typical Cash Flow [after Fanchi, 2001, Chapter 9]

The time dependence of NPV is illustrated in Figure 12-6. This figure shows that NPV can be negative. A negative NPV says that the investment option is operating at a loss. The loss is usually associated with initial capital investments and operating expenses that are incurred before the investment option begins to generate revenue. For example, mirrors have to be installed in a solar electric generating system and a generator built before electricity can be sold to the transmission grid. The installation of mirrors and a generator are capital expenses that occur before revenue is generated by the sale of electricity. Similarly, large investments in the design and construction of offshore platforms must be made before wells can be drilled and oil produced. The eventual growth in positive NPV occurs when revenue exceeds expenses. The point in time when NPV equals zero is the payout time. In the example shown in Figure 12-6, payout time is approximately 1.5 years. The concept of payout time can be applied to discounted cash flow or undiscounted cash flow.

Several commonly used indicators of economic performance are listed in Table 12-1. The discount rate at which the maximum value of NPV is zero is called the discounted cash flow return on investment

(DCFROI) or Internal Rate of Return (IRR). Payout time, NPV, and DCFROI account for the time value of money. An indicator that does not account for the time value of money is the profit-to-investment (PI) ratio. The PI is the total undiscounted cash flow without capital investment divided by total investment. It is often useful to prepare a variety of plots, such as NPV versus time and NPV versus discount rate, to show how economic indicators perform as functions of time.

Economic analyses using indicators of economic performance provide information about the relative performance of different investment options. The economic viability of an investment option is usually decided after considering a combination of economic indicators. For example, an investment option may be considered economically viable if the DCFROI is greater than 30% and payout time is less than two years. It should be remembered, however, that quantitative indicators provide useful information, but not complete information.

Table 12-1
Indicators of Economic Performance

Discount Rate	Factor to adjust the value of money to a base year.
Net Present Value	Value of cash flow at a specified discount rate.
DCFROI or IRR	Discount rate at which net present value is zero.
Payout Time	Time when net present value equals zero.
Profit-to-Investment Ratio	Undiscounted cash flow without capital investment divided by total investment.

Economic viability is influenced by both tangible and intangible factors. The tangible factors, such as building a generator or drilling a well, are relatively easy to quantify. Intangible factors such as environmental and socio-political concerns are relatively difficult to quantify, yet may be more important than tangible factors.

12.4.2 Life Cycle Analysis

The analysis of the costs associated with an energy source should take into account the initial capital expenditures and annual operating expenses for the life of the system. This analysis is called life cycle analysis and the costs are called life cycle costs. Life cycle costing requires the analysis of all direct and indirect costs associated with the system for the entire expected life of the system. According to Bent Sørensen [2000, page 762], life cycle analysis includes analyzing the impact of "materials or facilities used to manufacture tools and equipment for the process under study and it includes final disposal of equipment and materials, whether involving reuse, recycling or waste disposal." A list of life cycle costs is presented in Table 12-2.

Table 12-2
Life Cycle Costs of an Energy System
[Goswami, et al., 2000, page 528]

Capital equipment costs
Acquisition costs
Operating costs for fuels, etc.
Interest charges for borrowed capital
Maintenance, insurance, and miscellaneous charges
Taxes (local, state, federal)
Other recurring or one-time costs associated with the system
Salvage value (usually a cost) or abandonment cost

It is important to recognize that the future cost of some energy investment options may change significantly as a result of technological advances. The cost of a finite resource can be expected to increase as the availability of the resource declines, while the cost of an emerging technology will usually decline as the infrastructure for supporting the technology matures.

The initial costs of one energy system may be relatively low compared to competing systems. If we only consider initial cost in our analysis, we may adopt an energy option that is not optimum. For example, the annual operating expenses for an option we might choose based on initial cost may be significantly larger than those of an alternative option. In addition, projections of future cost may be substantially in error if the cost of one or more of the components contributing to an energy system changes significantly in relation to our original estimate. To avoid making less-than-optimum decisions, we should consider all of the life cycle costs of each investment option. We also need to evaluate the sensitivity of cash flow predictions to plausible changes in cost as a function of time.

Inherent in life cycle analysis is an accurate determination of end use efficiency. End use efficiency is the overall efficiency of converting primary energy to a useful form of energy. It should include an analysis of all factors that affect the application. As a simple example, consider the replacement of a light bulb. The simplest decision is to choose the least expensive light bulb. On a more sophisticated level, we need to recognize that the purpose of the light bulb is to provide light and some light bulbs can provide light longer than other light bulbs. In this case we need to consider the life of the light bulb in addition to its price. But there are still more factors to consider. If you live in an equatorial region, you might prefer a light bulb that emits light and relatively little heat, so you can reduce air conditioning expenses. On the other hand, if you live in a cooler northern climate, you might desire the extra heat and choose a light bulb that can also serve as a heat source. Once you select a light bulb, you want to use it where it will do the most good. Thus, if you chose a more expensive light bulb that has a long life and generates little heat, you would probably prefer to use the light bulb in a room where the bulb would be used frequently, such as a kitchen or office, rather than a closet where the bulb would be used less frequently. All of these factors should be taken into account in determining the end use efficiency associated with the decision to select a light bulb.

One of the goals of life cycle analysis is to make sure that decision makers in industry, government, and society in general, are aware of all of the costs associated with a system. In the context of energy resource management, Bent Sørensen [2000, Section 7.4.3] has identified the following impact areas: economic, environmental, social, security, resilience, development, and political. Some typical questions that must be answered include the following:

1. Does use of the resource have a positive social impact, that is, does resource use provide a product or service without adversely affecting health or work environment?
2. Is the resource secure, or safe, from misuse or terrorist attack?
3. Is the resource resilient, that is, is the resource relatively insensitive to system failure, management errors, or future changes in the way society assesses its impact?
4. Does the resource have a positive or negative impact on the development of a society, that is, does the resource facilitate the goals of a society, such as decentralization of energy generating facilities or satisfying basic human needs?
5. What are the political ramifications associated with the adoption of an energy resource?
6. Is the resource vulnerable to political instability or can the resource be used for political leverage?

A thorough life cycle analysis will provide answers to all of these questions. Of course, the validity of the answers will depend on our ability to accurately predict the future.

12.4.3 Risk Analysis and Real Options Analysis

A characteristic of natural resource management is the need to understand the role of uncertainty in decision making. The information we have about a natural resource is usually incomplete. What information we do have may contain errors. Despite the limitations in our knowledge, we must often make important decisions to advance a project. These decisions should be made with the recognition that risk, or uncertainty, is present and can influence investment decisions. Here, risk refers to the possibility that an unexpected event can adversely affect the value of an asset. Uncertainty is not the same as risk. Uncertainty is the concept that our limited knowledge and understanding of the future does not allow us to predict the consequences of our decisions with 100% accuracy. Risk analysis is an attempt to quantify the risks associated with investing under uncertainty.

One of the drawbacks of traditional risk analysis is the limited number of options that are considered. The focus in risk analysis is decision making based on current expectations about future events. For example, the net present value analysis discussed above requires forecasts of revenue and expenses based on today's expectations. Technological advances or political instabilities are examples of events that may significantly alter our expectations. We might overlook or ignore options that would have benefited from the unforeseen events. An option in this context is a set of policies or strategies for making current and future decisions. Real Options Analysis attempts to incorporate flexibility in the management of investment options that are subject to considerable future uncertainty.

The best way to incorporate options in the decision making process is to identify them during the early stages of analysis. Once a set of options has been identified for a particular project, we can begin to describe the uncertainties and decisions associated with the project. By identifying and considering an array of options, we obtain a more complete picture of what may happen as a consequence of the decisions we make. Real Options Analysis helps us understand how important components of a project, particularly components with an element of uncertainty, influence the value of the project.

Point to Ponder: Is the military budget a hidden cost of importing oil and gas?

Most industrialized nations rely on oil and gas to support their energy needs. Some of these countries, such as the United States, Great Britain, Russia and China maintain significant military capabilities for national defense and global influence. In the case of Great Britain and the United States, part of this military expenditure is used to maintain access to oil and gas resources. These costs should be considered part of the cost of continued reliance on fossil fuels. For example, if US$10 billion is spent to keep open supply lines to a region in one year, and you import 500 million barrels of oil from that region in a year, that would add US$40.00 to the cost of each barrel of imported oil, or about US$1.00 to each gallon of gasoline. This cost assumes that no lives are lost and, if they are, that they have no monetary value. Each society must determine the value they want to place on the lives of people who are being asked to protect their sources of energy (Figure 12-7).

Figure 12-7. Military Cemetery in Washington, D.C. (Fanchi, 2002)

12.5 Levelized Energy Cost

A useful measure for evaluating the feasibility of an energy generating technology and comparing different energy generating systems is levelized energy cost (LEC). LEC is an estimate of the cost of generating energy, typically in the form of electricity, for a particular energy generating system. The LEC includes all costs incurred over the expected life of the system. A net present value calculation is performed to calculate the break-even price for energy generated by the system. Levelized costs of electricity generation are illustrated in Table 12-3.

Table 12-3
Levelized Costs of Electricity Generation
[IER website, 2010]

	2007 US$/kWh	Normalized
Gas, Advanced CC	0.080	0.748
Gas, Combined Cycle	0.084	0.785
Coal, Conventional	0.095	0.888
Coal, Advanced	0.103	0.963
Nuclear	0.107	1.000
Biomass	0.107	1.000
Geothermal	0.112	1.047
Hydroelectric	0.114	1.065
Gas, Advanced CC+CCS	0.116	1.084
Coal, Advanced+CCS	0.123	1.150
Wind	0.141	1.318
Wind, Offshore	0.230	2.150
Solar Thermal	0.264	2.467
Solar PV	0.396	3.701

The second column of Table 12-3 presents average national levelized costs for electricity generating plants that rely on different energy sources. The levelized costs refer to plants that will enter service in 2016 and are from an analysis of the updated AEO 2009 reference case [IER website, 2010] by the Institute for Energy Research (IER). IER is a market-oriented non-profit organization which analyzed information from the 2009 Annual Energy Outlook (AEO) published by the United States Energy Information Administration. The acronym CC refers to combined cycle, and CCS refers to carbon capture and sequestration. In some cases, CCS refers to carbon capture and storage. The levelized costs in Table 12-3 are normalized in the third column using the levelized cost of a nuclear power plant as the reference cost for normalization.

The levelized costs in Table 12-3 include a 3% increase in the cost of capital for technologies that do not mitigate greenhouse gas emissions. The 3% increase presumes that the cost of a technology like a coal-fired power plant will have to increase to pay for technology to reduce greenhouse gas emissions or buy credits to allow the facility to operate without carbon capture and sequestration technology. This means the levelized capital costs of technologies like coal-fired power plants without CCS technology are higher than the actual cost of existing plants with the same technologies. The inflated capital cost is a penalty that is applied in anticipation of new government policies designed to regulate greenhouse gas emissions. The levelized costs do not include financial incentives such as tax breaks or government subsidies. More details about the levelized costs are provided by the IER [IER website, 2010].

Figures 12-8 and 12-9 graphically display the information from Table 12-3. The comparison shows that the levelized costs of some renewable energy systems are comparable to the levelized costs of some systems that consume fossil fuels, especially if the levelized cost of CCS technology is included. For example, wind power is a few cents per kWh more expensive than fossil fuels when expressed in 2007 US$. One way to make wind power more competitively economically is to add a carbon tax to combustible fuels, such as natural gas or coal, to account for the environmental impact of greenhouse gas emissions. A carbon tax that is the difference between the cost of wind power and the cost of fossil fuel power would only have to be a few cents per kWh to make wind power competitive with fossil fuel power.

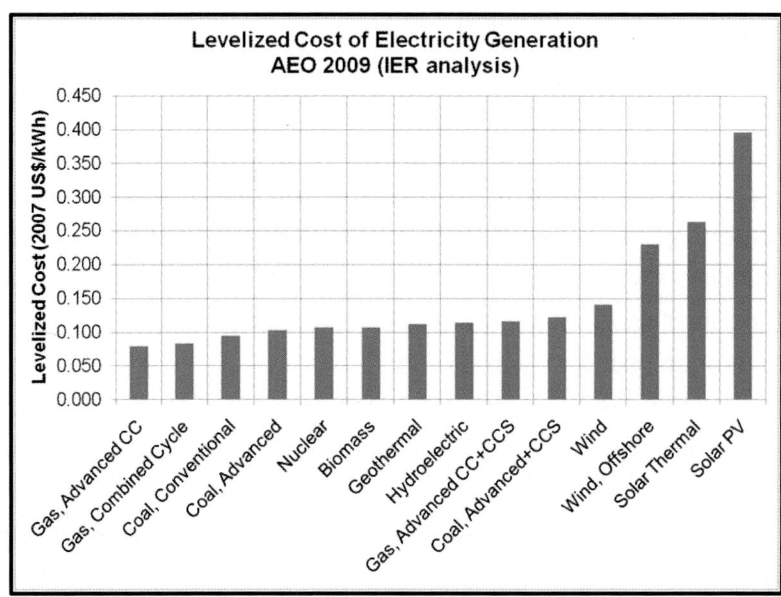

Figure 12-8. Comparison of Levelized Cost of Electricity Generation

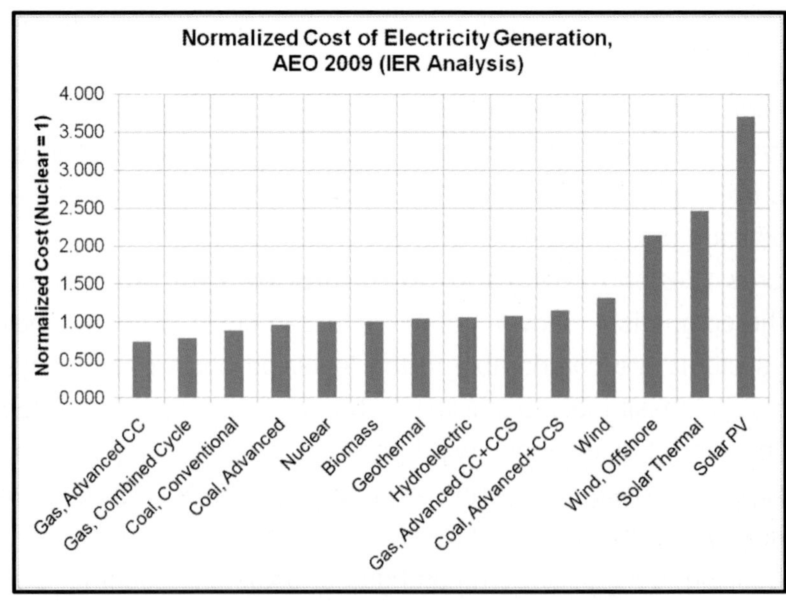

Figure 12-9. Comparison of Normalized Cost of Electricity Generation

Point to Ponder: Can we bias cost comparisons?
Every cost comparison relies on assumptions. It is possible to bias a comparison of costs by selecting assumptions that either increase or reduce the cost of a particular technology. For example, if we include the cost of carbon capture and sequestration (CCS) technology in the cost comparison, we can increase the cost of fossil fuel-fired power plants. On the other hand, we can reduce the cost of electricity generation using renewable energy technology by assuming that tax credits or subsidies will be applied. The bias can be magnified if we combine assumptions that favor one technology relative to another. If the bias is significant, it can change the ranking of technologies.

12.6 ACTIVITIES

True-False

Specify if each of the following statements is True (T) or False (F).

1. Microeconomics is the study of how individual businesses or households decide to allocate limited resources.
2. An increase in price leads to an increase in demand for a product.
3. Demand for oil in the United States is inelastic in the price range of \$40/bbl to \$80/bbl (in 2010 U.S. dollars).
4. OPEX for wind and solar power are zero.
5. Overhead is an indirect cost.
6. Electricity is fungible because one unit of electricity can be interchanged with another, regardless of the source.
7. The net present value of a project is always positive.
8. Salvage value can be either a revenue or a cost and should be included in a life cycle analysis.
9. Risk is the concept that our limited knowledge and understanding of the future does not allow us to predict the consequences of our decisions with 100% accuracy.
10. The goal of LEC is to compare energy costs across energy systems.

Questions

1. Use the following table to estimate the percent of world energy consumption that was due to fossil fuels in 2002.

Primary Energy Type	Total World Energy Consumption (Source: EIA Website, 2002)
Oil	39.9 %
Natural Gas	22.8 %
Coal	22.2 %
Hydroelectric	7.2 %
Nuclear	6.6 %
Geothermal, Solar, Wind and Wood	0.7 %

2. What is life cycle analysis?
3. What is the purpose of Real Options Analysis?
4. Suppose a nuclear power plant can provide electrical energy at a cost of 6.7 cents/kWh. The corresponding cost for pulverized coal is 4.2 cents/kWh and for natural gas averages 4.7 cents/kWh. If a carbon tax of 3 cents/kWh is imposed, will the cost of nuclear energy be more expensive or less expensive than the cost of energy from pulverized coal or natural gas?
5. Which renewable energy systems in Table 12-3 have a lower levelized cost than power plants that burn gas or coal and use CCS technology?
6. Explain the elasticity of demand.
7. What is the difference between OPEX and CAPEX?
8. Can intangible costs be quantified? Explain your answer.
9. In an economy of scale, the per-unit cost of producing a product decreases as the number of units produced (circle one) increases or decreases.
10. What is end use efficiency?

CHAPTER 13

FUTURE ISSUES – GEOPOLITICS OF ENERGY

Decisions are being made today by companies and governments around the world based on the assumption that oil will be replaced as the primary source of energy. Governments of oil consuming nations are maintaining military forces capable of keeping oil supply lines open between producers and consumers. Oil exporting nations are trying to optimize their revenues by influencing the market price of oil. How high will the price of oil go as supply dwindles? Will the price of oil rise without limit or will market factors constrain the price? Every one of us has a stake in the answers to these questions and the decisions that are being made to provide energy.

One of the problems facing society is the need to develop and implement a strategy that will provide energy to meet future global energy needs and satisfy environmental objectives. The development of strategies depends on our view of the future. At best, we can only make educated guesses about what the future will bring. The quality of our educated guesses depends on our understanding of the situation and the quantity and quality of information we have available. We can distinguish between different levels of guessing by giving different names to our predictions.

Figure 13-1 displays three different levels of predicting the future: stories, scenarios, and models. A story can be used to provide a qualitative picture of the future. Stories are relatively unclear because our understanding is limited and the information we have is relatively incomplete. As we gain information and understanding, we can begin to discuss scenarios. Scenarios let us consider different stories about complex situations. They let us incorporate more detail into plausible stories. Unlike forecasts, which let us extrapolate historical behavior to predict the future, scenarios let us consider the effects of discontinuities and sudden changes. Forecasts assume a certain degree of continuity from past to future, while the future may in fact be altered dramatically by an unexpected development. In the context of energy, revolutionary developments in nuclear fusion technology or unexpected cost reductions in solar energy technology could lead to abrupt changes in historical trends of energy production and consumption. These changes could invalidate any forecast that was based on a continuous extrapolation of historical trend and lead to a future that would have been considered implausible based on past performance.

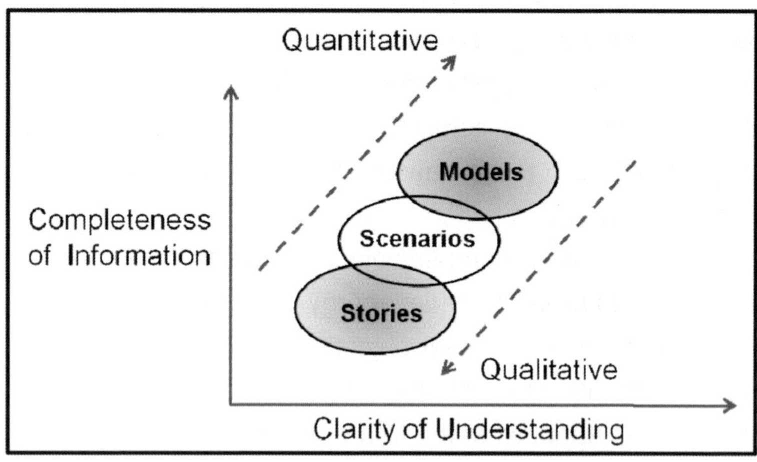

Figure 13-1. Stories, Scenarios and Models [after McKay, 2002]

We can construct models that allow us to quantify our scenarios as our knowledge and understanding increase. The concept of sustainable development is a road map of how we should prepare for the future and it is a vision of what the future should be. It is a scenario that has been adopted by the United Nations and helps explain evolving business practices in the energy industry. We discuss sustainable development as a road map for the future and present some forecasts that recognize the need to replace oil. Before we consider options to shape the future energy mix, we review how we arrived at this point, and why a change may be needed.

13.1 SUSTAINABLE DEVELOPMENT

An emerging energy mix is needed to meet energy demand in the 21st century. The demand for energy is driven by factors such as increasing trends in population and consumption. The ability to meet the demand for energy depends on such factors as price volatility, supply availability, and efficiency of energy use. One measure of how efficiently a country is using its energy is energy intensity. In the context of energy policy, energy intensity may be defined on the national level as the total domestic primary energy consumption divided by the gross domestic product. Countries that have low energy consumption and high domestic productivity will have relatively low energy intensities. Countries with high energy consumption and low domestic productivity will have relatively high energy intensities. By considering the change in energy intensity as a function of time, we can see if a country is improving its efficiency of energy consumption relative to its domestic productivity.

Figure 13-2 presents energy intensity (in BTU per Year 2005 U.S. dollars) as a function of time for a select group of countries. If one of our goals is to maintain or improve quality of life with improved energy efficiency, we would like to see the energy intensity of a nation decrease as a function of time. Energy intensity was increasing for Saudi Arabia up to 2007, and energy intensity was decreasing for the other countries in the sample, including the United States. China's energy intensity increased early in the 21st century and was declining by 2007. India's energy intensity increased in the latter part of the 20th century and was declining at the beginning of the 21st century.

The emerging energy mix is expected to rely on clean energy, that is, energy that is generated with minimal environmental impact. The goal is sustainable development which includes the consideration of the rights of future generations. The concept of rights is a legal and philosophical concept. It is possible to argue that future generations do not have any legal rights to current natural resources and are not entitled to rights. From this perspective, each generation must do the best it can with available resources. On the other hand, many societies are choosing to adopt the value of preserving natural resources for future generations. National parks are examples of natural resources that are being preserved.

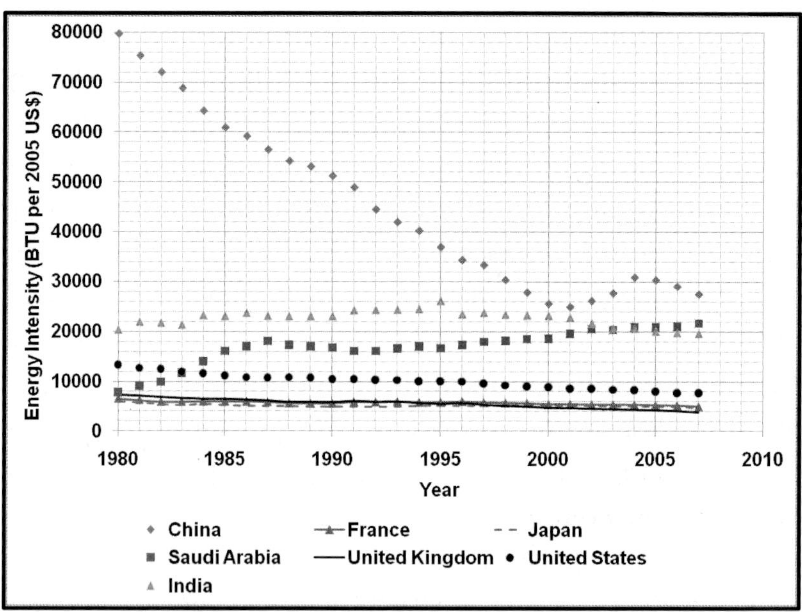

Figure 13-2. Energy Intensity for Selected Countries [US EIA website, 2010, International Energy Statistics]

One industry response to environmental and social concerns in the context of sustainable development is the 'triple bottom line' [Whittaker, 1999]. The three components of sustainable development and the three goals of the triple bottom line (TBL) are economic prosperity, social equity, and environmental protection. From a business perspective, the focus of TBL is the creation of long-term shareholder value by recognizing that corporations are dependent on licenses provided by society to do business. If business chooses not to comply with sustainable development policies, society can enforce compliance by imposing additional government regulation and control.

13.2 ENERGY AND ETHICS

One issue that must be considered in the context of sustainable development is the distribution of energy. Should energy be distributed around the world based on need, ability to pay, or some other value? This question is an ethical issue because the answer to the question depends on the values we choose to adopt.

The distribution of energy in the future will depend on whether or not a nation has a large per capita energy base. Should nations with energy resources help those in need? If so, how should they help? Traditional ethics would favor a policy of helping those nations without energy resources, but opinions differ on how to proceed. Two of the more important ethical positions are identified as *lifeboat ethics* and *spaceship ethics.* These positions are considered here for two reasons: they are diametrically opposed ethical positions that apply to the global distribution of energy; and they illustrate that people of good will can take opposite positions in a significant debate.

Proponents of *lifeboat ethics* oppose the transfer of wealth by charitable means. In this view, the more developed industrial nations are considered rich boats and the less developed, overcrowded nations are poor boats. The rich boats should not give the poor boats energy because their help would discourage the poor boats from making difficult choices such as population control and investment in infrastructure. Lifeboat ethics is a "tough love" position; it encourages nations to seek self-sufficiency. On the other hand, it might make some nations desperate and encourage the acquisition of energy resources by military means.

Proponents of *spaceship ethics* argue that everyone is a passenger on spaceship Earth. In this view, some passengers travel in first class while others are in steerage. A more equitable distribution of energy is needed because it is morally just and it will prevent revolts and social turmoil. Thus, the wealthy should transfer part of their resources to the poor for both moral and practical reasons. On the other hand, nations that receive charitable donations of energy may be unwilling to make the sacrifices needed to become self-sufficient.

It may be that a synthesis of these two ethical positions would be the optimum policy for distributing energy. Countries with resources could help the other countries to develop their resources with the clear understanding that the developing countries must become self-sufficient.

Point to Ponder: Do we need to consider the ethics of energy distribution?
The issue of energy distribution is really an issue of energy access. If people do not have access to essential resources, they may feel compelled to fight for those resources. Ethical positions such as *lifeboat ethics* and *spaceship ethics* attempt to address these concerns. Can access to energy be provided without discouraging self-reliance? Can the amount of useful energy be increased so that the need for sacrifice is lessened and the cost of energy is decreased? These are questions facing us today.

13.3 Energy and Geopolitics

Quality of life, energy, and the distribution of energy are important components of global politics. Readily available, reasonably priced energy is a critical contributor to the economic well-being of a nation. We have already seen that deforestation in England motivated the search for a new primary fuel. The need for oil encouraged Japanese expansion throughout Asia in the 1930's and was one of the causes of World War II. The 1973 Arab-Israeli war led to the first oil crisis with a short-term, but significant increase in the price of oil. This oil price shock was followed by another in 1979 after the fall of the Shah of Iran. These oil price increases are considered shocks because they were large enough to cause a significant decline in global economic activity [Verleger, 2000, page 76]. Our ability to correctly forecast energy demand depends on our understanding of technical and socio-political issues. In this section, we give a brief introduction to global politics and then discuss its implications for models of future energy demand.

13.3.1 Clash of Civilizations

The world has been undergoing a socio-political transition that began with the end of the Cold War and is continuing today. S.P. Huntington [1996] provided a view of this transition that helps clarify historical and current events, and provides a foundation for understanding the socio-political issues that will affect energy demand.

Huntington argued that a paradigm shift was occurring in the geopolitical arena. A paradigm is a model that is realistic enough to help us make predictions and understand events, but not so realistic that it tends to confuse rather than clarify issues. A paradigm shift is a change in paradigm. A geopolitical model has several purposes. It lets us order events and make general statements about reality. We can use the model to help us understand causal relationships between events and communities. The communities can range in size from organizations to alliances of nations. The geopolitical model lets us anticipate future developments and, in some instances, make predictions. It helps us establish the importance of information in relation to the model and it shows us paths that might help us reach our goals.

The Cold War between the Soviet Union and the Western alliance led by the United States established a framework that allowed people to better understand the relationships between nations following the end of World War II in 1945. When the Cold War ended with the fall of the Berlin wall and the break-up of the Soviet Union in the late 1980's and 1990's, it signaled the end of one paradigm and the need for a new paradigm. Several geopolitical models have been proposed. Huntington considered four possible paradigms for understanding the transition (Table 13-1).

Table 13-1
Huntington's Possible Geopolitical Paradigms

1	One Unified World
2	Two Worlds (West versus non-West; us versus them)
3	Anarchy (184+ Nation-states)
4	Chaos

The paradigms in Table 13-1 cover a wide range of possible geopolitical models. The One Unified World paradigm asserts that the end of the Cold War signaled the end of major conflicts and the beginning of a period of relative calm and stability. The Two Worlds paradigm views the world in an "us versus them" framework. The world was no longer divided by political ideology (democracy versus communism); it was divided by some other issue. Possible divisive issues include religion and rich versus poor (generally a North-South geographic division). The world could also be split into zones of peace and zones of turmoil. The third paradigm, Anarchy, views the world in terms of the interests of each nation, and considers the relationships between nations to be unconstrained. These three paradigms, One Unified World, Two Worlds, and Anarchy, range from simple (One Unified World) to complex (Anarchy).

The Chaos paradigm says that post-Cold War nations are losing their relevance as new loyalties emerge. In a world where information flows freely and quickly, people are forming allegiances based on shared traditions and value systems. The value systems are notably cultural and, on a more fundamental level, religious. The new allegiances are in many cases a rebirth of historical loyalties. New alliances are forming from the new allegiances and emerging as a small set of civilizations. The emerging civilizations are characterized by ancestry, language, religion, and way of life.

Huntington considered the fourth paradigm, Chaos, to be the most accurate picture of current events and recent trends. He argued that the politics of the modern world can be best understood in terms of a model that considers relationships between the major contemporary civilizations shown in Table 13-2. The existence of a distinct African civilization has been proposed by some scholars, but is not as widely accepted as the civilizations identified in the table.

Table 13-2
Huntington's Major Contemporary Civilizations

Civilization	Comments
Sinic	China and related cultures in Southeast Asia
Japanese	The distinct civilization that emerged from the Chinese civilization between 100 and 400 A.D.
Hindu	The peoples of the Indian subcontinent that share a Hindu heritage.
Islamic	A civilization that originated in the Arabian peninsula and now includes subcultures in Arabia, Turkey, Persia, and Malaysia.
Western	A civilization centered around the northern Atlantic that has a European heritage and includes peoples in Europe, North America, Australia, and New Zealand.
Orthodox	A civilization centered in Russia and distinguished from Western Civilization by its cultural heritage, including limited exposure to Western experiences (such as the Renaissance, the Reformation, and the Enlightenment).
Latin America	Peoples with a European and Roman Catholic heritage who have lived in authoritarian cultures in Mexico, Central America and South America.

Each major civilization has at least one core state [Huntington, 1996, Chapter 7]. France and Germany are core states in the European Union, which is viewed as part of Western Civilization. The United States is also a core state in Western Civilization. Russia and China are core states, perhaps the only core states, in Orthodox Civilization and Sinic Civilization, respectively. Core states are sources of order within their civilizations. Stable relations between core states can help provide order between civilizations.

The growth of multiculturalism in some states has established communities within those states that may not share the values and allegiances of the host state. A multicultural state in this context is a member state of one civilization that contains at least one relatively large group of people that is loyal to many of the values of a different civilization. For example, Spain is a member state of Western Civilization with a sizable Islamic population. After the collapse of the Soviet Union, some multicultural states (e.g. Yugoslavia and Czechoslovakia) that were once bound by strong central governments separated into smaller states with more homogeneous values.

Within the context of the multi-civilization geopolitical model, the two World Wars in the 20th century began as civil wars in Western Civilization and engulfed other civilizations as the hostilities expanded. Those wars demonstrate that civilizations are not monolithic: states within civilizations may compete with each other. Indeed, the growth of multiculturalism and large migrant populations in some states are making it possible for states within a civilization to change their cultural identity as cultures within member states compete for dominance. The change in cultural identity can lead to a change in allegiance to a civilization.

The Cold War and the oil crises in the latter half of the 20th century were conflicts between civilizations. Western Civilization has been the most powerful civilization for centuries, where power in this context refers to the ability to control and influence someone else's behavior. The trend in global politics is a decline in the political power of Western Civilization as other civilizations develop technologically and economically. Energy is a key factor in this model of global politics. This can be seen by analyzing the energy dependence and relative military strength of core states. For example, consider the relationship between Western and Islamic Civilizations.

Western Civilization is an importer of oil and many states in Islamic Civilization are oil exporters. The result is the transfer of wealth from oil importing states of Western Civilization to oil exporting states in Islamic Civilization. By contrast, the United States, a leading core state in Western Civilization, is the leading military power in the world with a large arsenal of nuclear weapons. Most core states in Islamic Civilization are relatively weak militarily and do not have nuclear weapons. The wealth being acquired by Islamic Civilization is being used to alter the balance of military power between Western Civilization and Islamic Civilization. Iran, a core state in Islamic Civilization, is using its oil wealth to improve its arsenal of conventional weapons and acquire nuclear technology from core states in other civilizations.

Point to Ponder: Does one country have the right to direct another country's technological development?

The United States, backed by United Nations resolutions, has attempted to block countries such as Iran and North Korea from developing nuclear weapons technologies. The goal was to prevent the proliferation of nuclear weapons, particularly to states that have governments that are deemed dangerous to Western Civilization. However, the policy of containing the distribution of nuclear technology also blocks countries from acquiring nuclear power which could help them improve quality of life for their people. This raises the question: should countries that have advanced technology be allowed, or even morally obligated, to control the distribution of technology to less-advanced societies? On the other hand, should a technologically advanced country be prohibited from sharing technology that can be used to build advanced weapons with a country that has shown signs of politically instability or militarily aggressiveness? Answers to these questions could lead to international agreements. Who is going to enforce these agreements? The United Nations seems like an obvious answer, but how effective has the United Nations been? Is it capable of enforcing agreements between nations, especially if powerful nations disagree on enforcement strategy and tactics?

Ideological differences between civilizations can lead to a struggle for global influence between core states. The battlefields in this struggle can range from economic to ideological to military. The outcome of this struggle depends on energy.

Energy importing states in one civilization rely on access to energy sources from energy exporting states in other civilizations. If the relationship between energy trading states is hostile, energy becomes a weapon in the struggle between civilizations. For example, the growth of non-Western civilizations, such as Sinic and Hindu Civilizations, has increased demand for a finite volume of oil. This increases the price of oil as a globally traded commodity and increases the flow of wealth between oil importing civilizations and oil exporting civilizations. Oil importing nations may try to reduce their need for imported oil by finding energy substitutes or by conservation. The social acceptability of energy conservation varies widely around the world. In some countries such as Germany, energy conservationists and environmentalists are a political force (the Green Party). In other countries, such as the United States, people may espouse conservation measures but be unwilling to participate in or pay for energy conservation practices, such as recycling or driving energy-efficient vehicles.

Energy production depends on the ability of energy producers to have access to natural resources. Access depends on the nature of relationships between civilizations with the technology to develop natural resources and civilizations with territorial jurisdiction over the natural resources. Much oil production technology was developed in Western Civilization and gave Western Civilization the energy it needed to become the most powerful civilization in the world. As Western Civilization consumed its supply of oil, it became reliant on other civilizations to provide it with the energy its states needed to maintain their oil-dependent economies. This dependence in times of stress between civilizations can lead to social turmoil and conflict between states that are members of different civilizations. Impending turmoil is motivating member states to develop alternatives to fossil fuels in an effort to achieve energy independence.

13.3.2 Clash Over Resources

Huntington's Clash of Civilizations provides one perspective on modern geopolitical events. Another perspective has been advocated by Michael T. Klare [2004], who argued that modern geopolitics is driven by a struggle for resources. Klare presented his thesis in his Preface [Klare, 2004, page xii]: "After examining a number of recent wars in Africa and Asia, I came to a conclusion radically different from Huntington's: that resources, not differences in civilizations or identities, are at the root of most contemporary conflicts." In **Resource Wars** [Klare, 2001], Klare made the case that oil, water, land and minerals were each important enough to be a source of contention, but he narrowed the list to petroleum in **Blood and Oil** [Klare, 2004]. Inexpensive and abundant petroleum is essential to modern economies, especially in the United States where energy based on fossil fuels supports the American lifestyle.

13.3.3 Energy Interdependence

Some people contend that the world can continue to rely on fossil fuels for decades to come. Robin Mills, petroleum economics manager at Emirates National Oil Company in Dubai, said that peak oil has not yet been reached and argues in the *Journal of Petroleum Technology* [Mills, 2009, page 17] that "approximately 3 billion barrels of ultimate conventional oil recovery (with approximately 1 billion produced so far) is a realistic minimum, compared to the peak oilers' 2 billion." The main cause for concern about oil supply is that oil prices have been too low to encourage exploration.

Oil and gas production technology has improved significantly since the first oil crisis in 1973. For example, seismic resolution has improved our image of the subsurface, multiple seismic surveys allow us to see how fluids move and where commercial deposits of oil and gas may be, completion techniques such as hydraulic fracturing have made resources in low permeability formations producible, and extended reach drilling of multilateral wells reduces the footprint and environmental impact of drilling. These technologies can be economically justified when the price of oil and gas are high enough, but the price needs to be sustained long enough to make a project commercially feasible.

The search for and development of new fossil fuel resources can be risky and expensive. Another source of resources is already well known: fields that have already been produced and abandoned often contain between one third and two thirds of their original resources. New techniques and higher prices may make many of these old resources commercial. Mills concluded that the "best place to look for new oil is in old fields, with fresh ideas" [Mills, 2009, page 17].

Mills notes that the negative public perception of the oil industry and claims that peak oil is imminent are increasing political pressure to prematurely shift away from fossil fuels. Furthermore, possible members of the next generation of oil industry professionals are choosing alternate career paths rather than entering a "sunset industry." The lack of interest in the oil industry from top graduates has been an industry concern for decades. For example, oil prices have historically tended to rise and fall in cycles. During one cycle, low oil prices in the 1980s led corporations to reduce their workforces with layoffs, which led to a perception that corporate employees were expendable.

Another issue to consider is the argument that combustion of fossil fuels causes so much environmental damage that it must be stopped. While methods of decreasing the environmental impact of burning coal, oil and natural gas are all improving, such as carbon capture and sequestration, these methods are being rejected or ignored by opponents of fossil fuels. People who seek to increase the cost of fossil fuels by imposing a tax on its use have influenced some governments. They argue the cost of using fossil fuels does not include the cost to mitigate environmental damage, and so the cost of fossil fuel combustion is understated and must be increased. This would make renewable energy options more attractive economically.

Mills suggested that the best way to achieve global energy security is for countries to recognize that a large energy-consuming nation would probably be unable to achieve energy independence at an acceptable cost. Instead, Mills argued that "energy security can only be achieved, or at least improved, by a balance between needs of exporters and importers, and a web of mutual interdependency" [Mills, 2009, page 17].

13.4 GLOBAL REGULATION OF CARBON EMISSIONS

Many countries have been evaluating different methods for minimizing carbon emissions into the atmosphere since the Kyoto Protocol was signed in 1997. The primary motivation for reducing greenhouse gas emissions is to control climate change. Several policies have been suggested for meeting carbon emissions goals set by the Kyoto Protocol. The leading option in the United States and European Union (EU) is called "Cap and Trade."

The premise behind Cap and Trade is that the central government of a country will be given a carbon emission goal by an international organization such as the United Nations. For example, European Union countries subject to the Kyoto Protocol agreed to reduce their carbon emissions by 8% relative to 1990 output levels by the year 2012. This is the "cap" for the country. The country can decide how to implement the policy. One implementation strategy proceeds as follows. The government determines the amount of emissions that can be produced and gives companies, including utilities that run heavily-emitting power plants, a certain amount of carbon credits that they can use. This is their allowed amount of carbon emissions. If a company emits less gas than it is allowed by its credits, the company may "trade" the unused credits to other companies that exceed their cap. Trade, in this case, usually means selling credits to the highest bidder.

The EU instituted a Cap and Trade program in January 2005. The Cap and Trade program, called the Emission Trading Scheme (ETS), began as a trial. In 2007, ETS progressed from the Phase I trial stage to

Phase II. Phase II links the EU ETS to that of other countries involved in similar carbon trading plans in accordance with the Kyoto Protocol.

The United States is not a signatory of the Kyoto Protocol because the Protocol was not ratified by the United States Senate. The primary concern of many Senators and critics of the Protocol was that the Protocol lacked provisions to control carbon emissions by China and India, which are the two most populous countries in the world. China and India have rapidly growing economies and an increasing demand for fossil fuels.

The Cap and Trade system can be designed to allow for the cap set on the country as a whole to be phased in. In other words, the country will reduce the number of carbon credits available each year to force a reduction in carbon emissions. The idea is that a gradual reduction would not require drastic, costly changes in a short period of time. Proponents of Cap and Trade point out that the system minimizes government intervention in the free market since companies may buy and sell credits at will and the market sets the price. The only responsibility the government has is to set the number of credits available each year. Critics of Cap and Trade express concern that Cap and Trade will not be effective at reducing greenhouse gas emissions, and that Cap and Trade is amenable to corruption.

13.4.1 Effect of Cap and Trade on Energy Supply and Demand

An example of a supply and demand curve offered by proponents of Cap and Trade is given in Figure 13-3. P_{Base} and Q_{Base} are the base price and quantity at equilibrium before any added legislation. In the figure, energy supply decreases due to the added cost of carbon-emitting products, but energy demand also decreases as consumers "do their part" to limit carbon emissions by decreasing energy use and as consumers change their purchasing habits because of the increased price of carbon-emitting products. The overall result is that prices remain stable and the amount of carbon-emitting products on the market decreases to $Q_{C\&T}$. The result is a win-win as consumer prices remain the same, but carbon emissions decrease.

Figure 13-4 implies that consumer behavior will not change with the additional cost of carbon emissions. While this may be overly simplified, it serves to illustrate our point that the quantity of energy that will be demanded will only decrease a small amount, but the price of energy will increase a substantial amount.

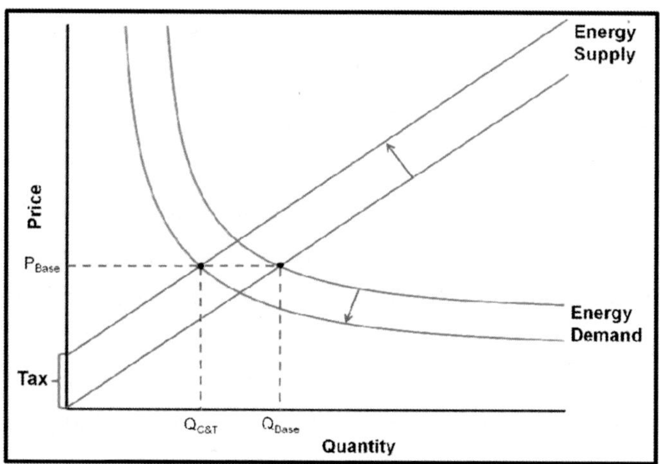

Figure 13-3. Proponents of Cap and Trade

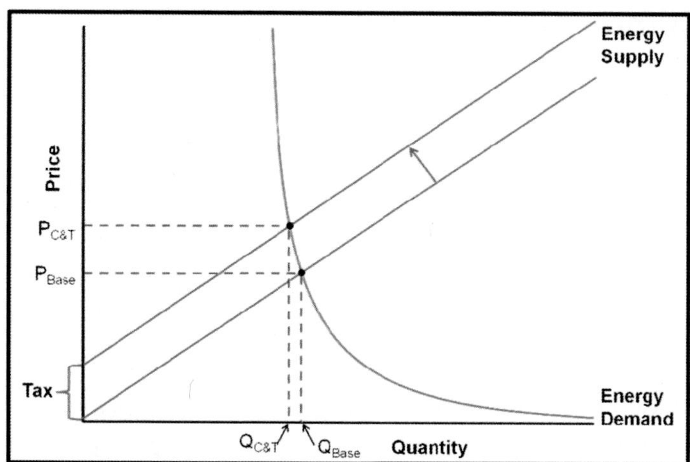

Figure 13-4. Opponents of Cap and Trade

Under Cap and Trade, energy providers are forced to buy carbon credits and bear the direct burden of the increase in cost. To the extent possible, the energy providers will shift the cost of carbon credits to consumers through price increases on consumer goods such as gasoline and electricity. This is why opponents of Cap and Trade programs argue that Cap and Trade programs are a tax on the middle and lower classes, that is, on people that spend the largest percentage of their income on energy.

We also need to consider the opportunity costs of Cap and Trade programs. If financial resources within a nation are being disproportionately spent to decrease carbon emissions, those same financial resources cannot be used for other purposes. This has a major impact on energy companies. While they can pass much of the cost on to consumers, they are not likely to be able to maintain the profit margins that existed prior to the Cap and Trade going into effect. Critics of the energy industry may argue that the industry does not need as much profit as it has made historically, but critics of Cap and Trade argue that the reduction in profit also means less money available for exploration, research, and projects that are risky and expensive. In this sense, it can be argued that Cap and Trade will tax everyone in the economy, including the poor, and it will stifle technological progress.

13.4.2 Climate Change and Adaptation

Some people believe that anthropogenic carbon is having a major impact on climate, while others believe the impact will not be significant. Still others believe that climate change has already begun and that it is too late to do anything about it.

James Lovelock introduced the earth feedback hypothesis in the 1960s. He hypothesized that the biosphere, which includes humanity, interacts with the earth's physical environment to maintain an environment that can sustain life. The earth feedback hypothesis was later dubbed the Gaia hypothesis after Gaia, the Greek supreme goddess of earth. Global climate change could be expected to alter the interaction between the biosphere and the physical components of the earth.

In **The Vanishing Face of Gaia** [Lovelock, 2009, page 68], Lovelock questioned the validity of climate forecasts and said the most important question in climate change is "How much and how fast is the earth heating?" Lovelock did not deny that climate change was occurring. On the contrary, Lovelock said that sea level change was a trustworthy indicator of the earth's heat balance. He pointed out that sea level was rising because ice was melting on land and the ocean was expanding because it was getting warmer. Evidence of ice melting is represented by the retreat and disappearance of glaciers. Figure 13-5 shows water flowing from a melting glacier near Saas Fe, Switzerland in June 2004. According to Lovelock, humans must accept that climate change is inevitable and prepare to adapt. He said that "Until we know for certain how to cure global heating, our greatest efforts should go into adaptation, to preparing those parts of the earth least likely to be affected by adverse climate change as the safe havens for a civilized humanity." [Lovelock, 2009, page 68]

Figure 13-5. Water Flowing from a Melting Glacier Near Saas Fe, Switzerland (Fanchi, 2004)

13.5 Activities

True-False

Specify if each of the following statements is True (T) or False (F).

1. Sustainable development is a policy that encourages society to meet its current needs without regard for future needs.
2. A goal of lifeboat ethics in the global distribution of energy is to encourage nations to seek energy self-sufficiency.
3. Political power is the ability to control and influence someone else's behavior.
4. A paradigm shift is a change in paradigm from one model to another.
5. Scenarios are the most quantitative method of predicting the future.
6. Huntington considered the end of the Cold War the catalyst for a geopolitical paradigm shift.
7. In Huntington's model, core states are sources of order within their civilizations.
8. Many oil fields that have already been abandoned still contain viable, recoverable oil.
9. Cap and Trade is amenable to corruption.
10. Cap and Trade systems could be considered a tax on the lower and middle classes.

Questions

1. Define spaceship ethics in the context of the global distribution of energy.
2. Why would wealthy nations be willing to implement spaceship ethics when developing policies for distributing energy?
3. Will the demand for energy depend more on the growth of energy demand in developed nations such as the United States and Germany or in developing nations such as India and China?
4. What is sustainable development?
5. What is energy intensity on the national level?
6. List Huntington's four potential geopolitical paradigms.
7. How has the rise of multiculturalism affected civilizations in Huntington's paradigm?

8. How does Klare's perspective on modern geopolitics differ from Huntington's?
9. Describe the concept behind Cap and Trade.
10. What is the Gaia hypothesis?

CHAPTER 14

FUTURE ISSUES – ENERGY FORECASTS

An energy mix is emerging to meet anticipated 21st century energy demand. Change in the energy mix is driven by factors such as increasing global population and energy consumption, the finite availability of fossil fuels, and climate change associated with industrialized society. As discussed in Chapter 13, the ability to meet demand for energy depends on such factors as energy density, price volatility, supply availability, and efficiency of energy use.

The energy types that contributed most to the energy mix in the latter half of the 20th century were wood, coal, oil, natural gas, water and nuclear. The emerging energy mix includes renewable and non-renewable energy resources. Increasing political pressure is causing the energy mix to trend toward renewable energy systems, but technological and commercial factors are justifying the inclusion of non-renewable energy systems, even if only in a more complementary capacity.

Future energy demand is expected to grow substantially as global population increases and developing nations seek a higher quality of life. Forecasts of the 21st century energy mix show that a range of scenarios is possible. One way to make a forecast is to estimate the size of the population in the future, and specify a desirable quality of life. We presented a correlation in Chapter 1 between energy consumption and quality of life as expressed by the United Nations Human Development Index. Combining population and per capita energy consumption establish energy demand. The next step is to assume technology is capable of supplying the energy needed to meet demand. The ability to meet demand is feasible if economics is not allowed to constrain development.

Forecasts of energy production depend on the ability of energy producers to have access to natural resources. Access depends, in turn, on the nature of relationships between civilizations with the technology to develop natural resources and civilizations with territorial jurisdiction over natural resources. Nations are facing energy challenges that require important policy choices: should they seek energy independence or interdependence? Should they be concerned about environmental sustainability? How can they optimize economic opportunity? A selection of energy forecasts is presented here to illustrate the range of options each of us should consider as decision-makers.

The United States Energy Information Administration presents energy supply and demand forecasts to 2030. This is a relatively short-term forecast that extrapolates existing technology. Our focus here is on long-term forecasts that cover all of the 21st century. The forecasts vary from a scenario that relies on nuclear energy to a scenario that relies on renewable energy. Other forecasts of the 21st century energy mix show a gradual transition from the current dependence on carbon-based fuels to a more balanced dependence on a variety of energy sources.

14.1 Nuclear Energy Forecast

P.E. Hodgson [1999] presented a scenario in which the world would come to rely on nuclear fission energy. He defined five Objective Criteria for evaluating each type of energy: capacity, cost, safety, reliability, and effect on the environment. The capacity criterion considered the ability of the energy source to meet future energy needs. The cost criterion considered all costs associated with an energy source. The safety criterion examined all safety factors involved in the practical application of an energy source. This includes hazards associated with manufacturing and operations. The reliability criterion considered the availability of an energy source. By applying the five Objective Criteria, Hodgson concluded that nuclear fission energy was the most viable technology for providing global energy in the future.

According to Hodgson, nuclear fission energy is a proven technology that does not emit significant amounts of greenhouse gases. He argued that nuclear fission reactors have an exemplary safety record when compared in detail with other energy sources. Breeder reactors could provide the fuel needed by nuclear fission power plants and nuclear waste could be stored in geological traps. The security of nuclear power plants in countries around the world would be assured by an international agency such as the United Nations. In this nuclear scenario, renewable energy sources would be used to supplement fission power and fossil energy use would be minimized. Hodgson did not assume that the problems associated with nuclear fusion would be overcome. If they are, nuclear fusion could also be incorporated into the energy mix.

Point to Ponder: What has hindered the global adoption of nuclear energy?

Nuclear energy is a long-term source of abundant energy that has environmental advantages. The routine operation of a nuclear power plant does not produce gaseous pollutants or greenhouse gases like carbon dioxide and methane. Despite its apparent strengths, the growth of the nuclear industry in many countries has been stalled by the public perception of nuclear energy as a dangerous and environmentally undesirable source of energy. This perception began with the use of nuclear energy as a weapon (Figure 14-1) and has been reinforced by widely publicized accidents at two nuclear power plants: Three Mile Island, Pennsylvania and Chernobyl, Ukraine. There are significant environmental and safety issues associated with nuclear energy, particularly nuclear fission. From an environmental perspective, the benefit of not producing greenhouse gases may be offset by the highly toxic and, in some cases, long-lived radioactive waste created by fission.

Figure 14-1. "Mushroom Cloud" Associated with the Detonation of a Nuclear Weapon

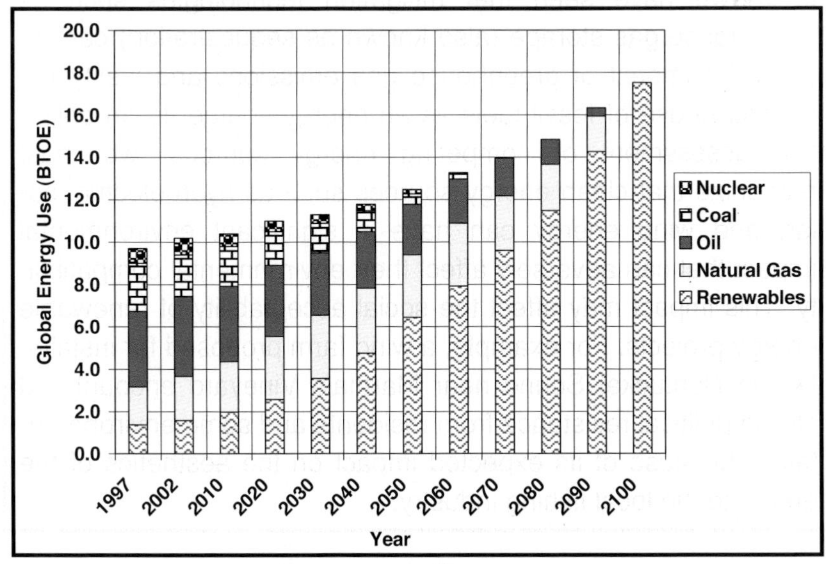

Figure 14-2. Forecast of a 21st Century Energy Mix that Eventually Relies on Renewable Energy Only [after Geller, 2003, page 227]

14.2 Renewable Energy Forecast

H. Geller [2003] presented a renewable energy scenario that sought to replace both nuclear energy and fossil energy with renewable energy only. An important objective of his forecast was to reduce greenhouse gas emissions to levels that are considered safe by the Kyoto Protocol. Figure 14-2 summarizes Geller's energy forecast. The figure shows global energy use as a function of time. Global energy use is expressed in billions of tons of oil equivalent (BTOE).

> **Point to Ponder: Is the environmental impact of clean energy sources an important consideration?**
> Society is searching for environmentally compatible, reliable energy sources. One of the characteristics of an environmentally compatible energy source is its cleanliness. A clean energy source emits negligible amounts of greenhouse gases or other pollutants. Even though fossil fuels have serious pollution problems, we have seen that mitigation technologies such as greenhouse gas storage (also known as sequestration) can reduce the impact of greenhouse gas emissions and justify the continued use of fossil fuels as an energy source. In an objective assessment of competing energy sources, we must recognize that clean energy sources such as hydroelectric, solar, and wind energy can have a significant environmental impact that can adversely affect their environmental compatibility. This impact may affect the social acceptability of renewable energy projects. For example, a wind farm proposed for installation in Nantucket Sound near Martha's Vineyard encountered strong political resistance from residents and some environmentalists because of its expected impact on the aesthetics of the area and the local fishing industry.

14.3 Energy Conservation and Energy Forecasts

Forecasts require an assumption about the role of energy conservation. Energy can be conserved using a number of very simple techniques. Some simple energy conservation methods that each of us can adopt include walking more and driving less; carpooling; planning a route that will let you drive to all of your destinations in the shortest distance; and using passive solar energy to dry your clothes or heat your home. Other energy conservation methods are more complicated: they are designed to reduce energy losses by improving the energy conversion efficiency of energy consuming devices. For example, improving gas mileage by reducing the weight of a vehicle or increasing the efficiency of internal combustion engines can reduce energy loss. We can decrease the demand for energy consuming activities such as air conditioning in the summer or heating in the winter by developing and using more effective insulating materials.

Energy conservation may be improved by increasing the efficiency of converting energy from one form into another. This efficiency is called energy conversion efficiency. It is the ratio of energy output to energy input. If we can decrease energy lost or wasted by a system, we can increase energy conversion efficiency. We must keep in mind that converting energy from its primary form to end-use energy will consume some energy and generate some waste, so it is unrealistic to expect to achieve 100% energy conversion efficiency.

One method for improving energy conversion efficiency is to find a way to use energy that would otherwise be lost as heat. Cogeneration is the simultaneous production and application of two or more sources of energy. The most common example of cogeneration is the simultaneous generation of electricity and useful heat. In this case, a fuel like natural gas can be burned in a boiler to produce steam. The steam drives an electric generator and is recaptured for such purposes as heating or manufacturing. Cogeneration is most effective when the cogeneration facility is near the site where excess heat can be used. The primary objective of cogeneration is to reduce the loss of energy by converting part of the energy loss to an energy output.

Energy conservation carries varying levels of importance in different countries. Some governments, especially in energy importing nations, are encouraging or requiring the development of energy conserving technologies. We can expect energy conservation to increase in the future as a result of more widespread adoption of energy conservation measures and by improvements in energy conversion efficiency. However, we should not expect energy conservation to be enough to satisfy global energy needs.

14.4 Energy Mix Forecasts

The forecasts mentioned in Section 14.1 and 14.2 focus on a future in which a single energy type (nuclear) or group of energy types (renewables) become the only energy forms used. The forecast discussed here is based on W.E. Schollnberger's 2006 forecasts, which cover the entire 21st century and predict the contribution of a variety of energy sources to the 21st century energy portfolio. We do not consider Schollnberger's forecast because it is right. We already know that it was in error within four years of its publication date. Schollnberger's forecast is worth studying because it illustrates how to use more than one scenario to project energy consumption for the entire 21st century.

Schollnberger [1999] considered three forecast scenarios:

A. "Another Century of Oil and Gas" corresponding to continued high hydrocarbon demand;

B. "The End of the Internal Combustion Engine" corresponding to a low hydrocarbon demand scenario; and

C. "Energy Mix" corresponding to a scenario with intermediate demand for hydrocarbons and an increasing demand for alternative energy sources.

Schollnberger viewed Scenario C as the most likely scenario. It is consistent with the observation that the transition from one energy source to another has historically taken several generations. Leaders of the international energy industry have expressed a similar view that the energy mix is undergoing a shift from liquid fossil fuels to other fuel sources.

There are circumstances in which Scenarios A and B could be more likely than Scenario C. For example, Scenario B would be more likely if environmental issues led to political restrictions on the use of hydrocarbons and an increased reliance on conservation. Scenario B would also be more likely if the development of a commercially competitive fuel cell for powering vehicles reduced the demand for hydrocarbons as a transportation fuel source. Failure to develop alternative technologies would make Scenario A more likely. It assumes that enough hydrocarbons will be supplied to meet demand.

Scenario C shows that natural gas will gain in importance as the economy shifts from a reliance on hydrocarbon liquid to a reliance on hydrocarbon gas. Eventually, renewable energy sources such as biomass and solar will displace oil and gas (Figure 14-3).

The demand by society for petroleum fuels should continue at or above current levels for a number of years, but the trend seems clear (see Figure 14-4). The global energy portfolio is undergoing a transition from an energy portfolio dominated by fossil fuels to an energy portfolio that includes a range of fuel types. Schollnberger's Scenario C represents one possible energy portfolio and the historical and projected energy consumptions trends are illustrated in Figure 14-4.

Schollnberger updated his energy forecast model in 2006 [Schollnberger, 2006]. He identified three desires that should influence energy decisions: sustained strong economic growth, security of energy supply, and a clean and safe environment. He argued that demand would determine future energy mix. Schollnberger reminded his European readers that many of them or their ancestors survived war-ravaged Europe at the end of World War II when the energy infrastructure was in ruins. Six decades later Europe has a modern energy infrastructure. Schollnberger, a champion of technology, was optimistic that humanity would achieve a sustainable energy future.

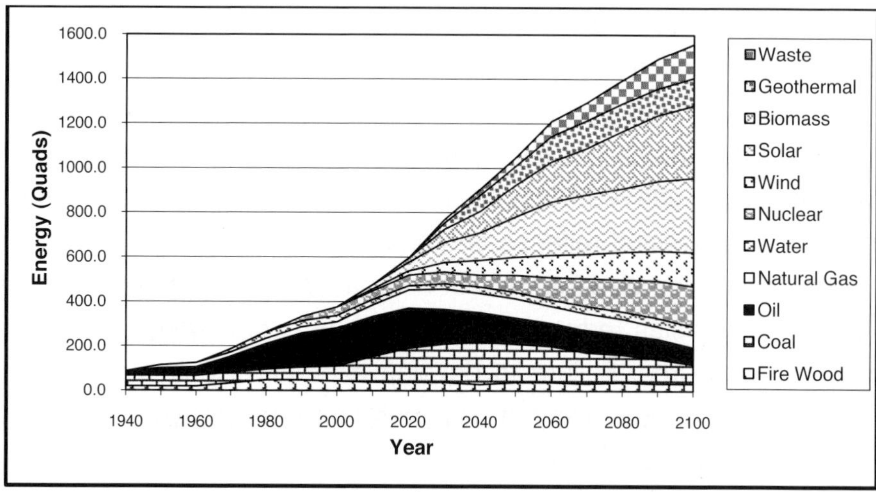

Figure 14-3. Forecast of 21st Century Energy Consumption

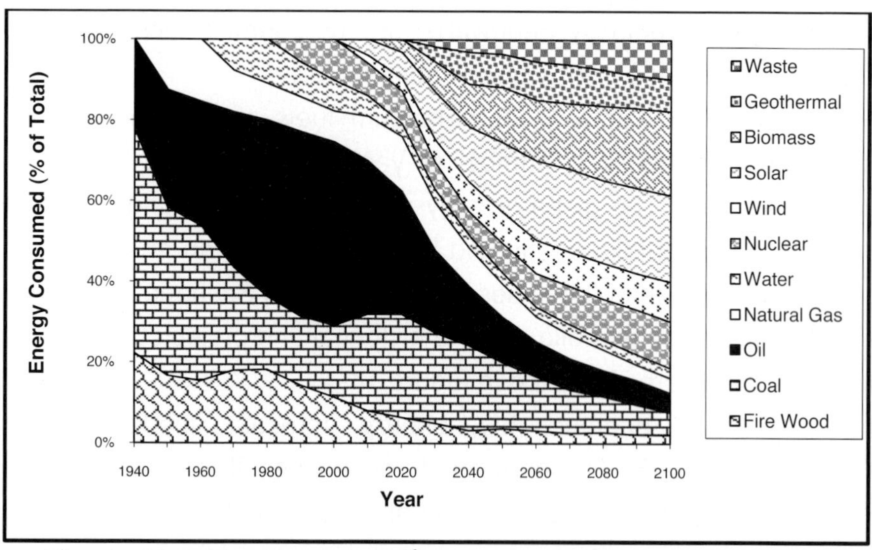

Figure 14-4. Forecast of 21st Century Energy Consumption as % of Energy Consumed

14.5 FORECASTS BASED ON SUPPLY

Schollnberger's forecast is based on demand. An alternative approach is to base the forecast on supply. Beginning with M.K. Hubbert [1956], several authors have noted that annual U.S. and world oil production approximately follows a bell shaped (Gaussian) curve. Forecasts based on Gaussian fits to historical data can be readily checked using publicly available data.

Figure 14-5 shows a Gaussian curve fit of world daily oil production data through year 2000, while Figure 14-6 shows a match of daily oil production through year 2008. The data is from the United States Energy Information Administration.

The fit of data in both figures is designed to match the most recent part of the production curve most accurately. This gives a match that is similar to results obtained by K.S. Deffeyes [2001, page 147]. The peak oil production rate in Figure 14-5 below occurs in 2010 and cumulative oil production by year 2100 is a little less than 2.1 trillion barrels.

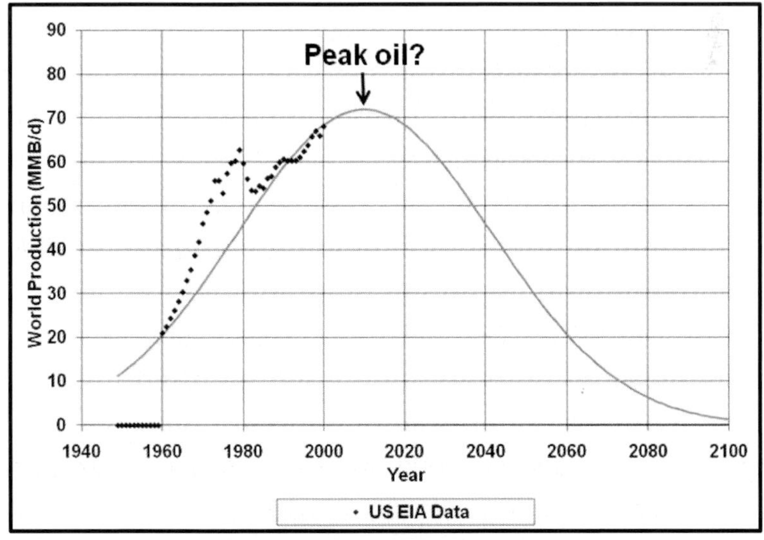

Figure 14-5. Oil Forecast by Matching Daily Oil Produced Through Year 2000 Using Gaussian Curve

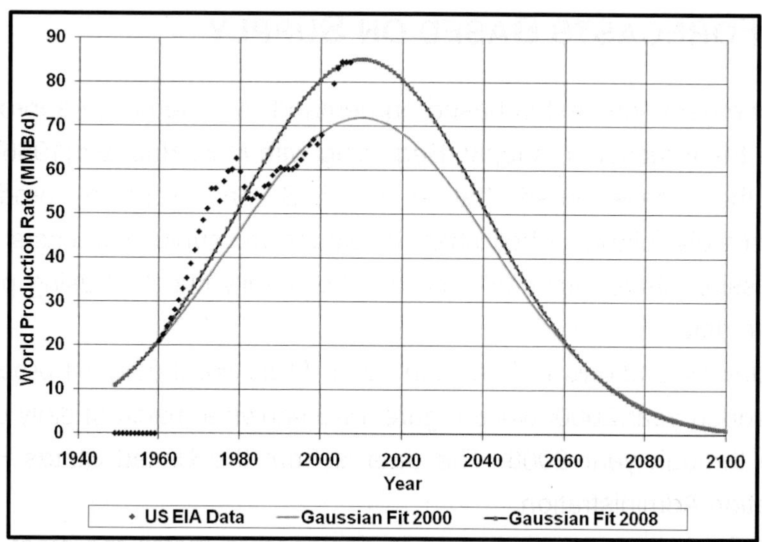

Figure 14-6. Oil Forecast by Matching Daily Oil Produced Through Year 2008 Using Gaussian Curve

Daily oil production through year 2008 shown in Figure 14-6 shows that oil production rate increased significantly between 2000 and 2008. The increase represents an increase in Saudi Arabian oil production capacity, which was motivated by the abrupt rise in oil price between 2006 and 2008. This rate increase shows that world oil production rate is constrained in part by infrastructure capability, but also by demand and price. Decline in oil price and production rate after 2008 is related to a global recession that began in 2008 and an associated decline in demand. This example illustrates the complexity of factors that influence fossil fuel production and consumption.

Analyses of historical data using a Gaussian curve typically predict that world oil production will peak in the first or second decade of the 21st century. By integrating the area under the Gaussian curve, forecasters have claimed that cumulative world oil production will range from 1.8 to 2.1 trillion barrels. These forecasts usually underestimate the sensitivity of oil production to technical advances and price. In addition, forecasts often discount the large volume of oil that has been discovered but not yet produced because the cost of production has been too high.

If we accept a Gaussian fit of historical data as a reasonable method for projecting oil production, we can estimate future oil production rate as a percentage of oil production rate in a specific base year. Figure 14-7 shows this estimate for base year 2000. According to this approach, world oil production rate will decline to 50% of year 2000 world oil production by the middle of the 21st century. For comparison, consider Schollnberger's Scenario C.

Differences between forecasts and actual data can be used to test the validity of assumptions that are used to generate forecasts. A test of forecast validity is the peak of world oil production. Forecasts of world oil production peak tend to shift as more historical data is accumulated. J.H. Laherrère [2000] pointed out that curve fits of historical data should be applied to activity that is "unaffected by political or significant economic interference, to areas having a large number of fields, and to areas of unfettered activity" (pg. 75). Furthermore, curve fit forecasts work best when the inflection point (or peak) has been passed. It does not appear that the inflection point has been reached, which increase the speculative character of forecasts.

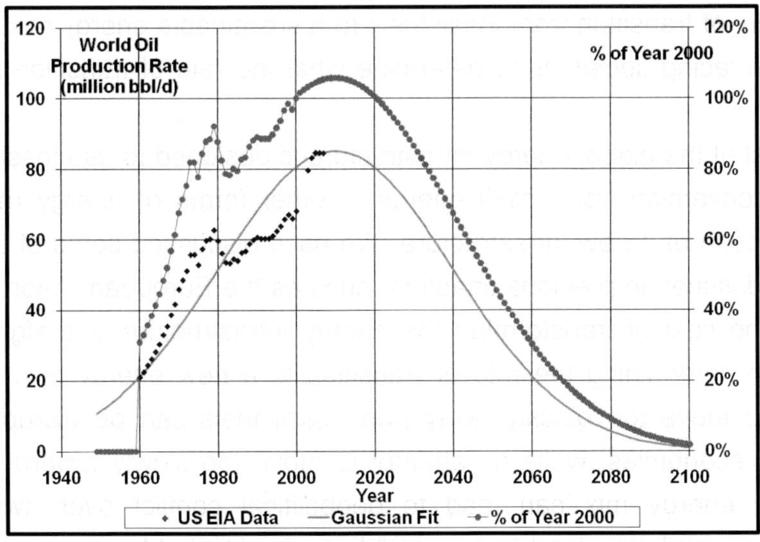

Figure 14-7. Forecast of 21st Century Oil and Gas Consumption as % of Oil and Gas Consumed in Year 2000

14.6 THE FUTURE OF ENERGY

The validity of an energy forecast depends on technical feasibility, economic viability, and political decisions. Social concern about nuclear waste and proliferation of nuclear weapons is a significant deterrent to reliance on nuclear fission power. These concerns are alleviated to some extent by the safety record of the modern nuclear power industry. Society's inability to resolve the issues associated with nuclear fusion make fusion an unlikely contributor to the energy mix until at least the middle of the 21st century. If nuclear fusion power is allowed to develop and eventually becomes commercially viable, it could become the primary energy source. Until then, the feasible sources of energy for use in the future energy mix are fossil fuels, nuclear fission, and renewable energy.

Concerns about greenhouse gas emissions, climate change, and security of the energy supply are encouraging a movement away from fossil fuels. Political instability in countries that export oil and gas, the finite size of known oil and gas supplies, increases in fossil fuel prices and decreases in renewable energy costs, such as wind energy costs, are motivating the adoption of renewable energy. It appears that the 21st century will be a century of transition from fossil fuels to a sustainable energy mix. A key decision facing society is to determine what the rate of transition should be.

Most of the global energy infrastructure is designed to use fossil energy. A conversion from fossil energy to other forms of energy requires construction of a new infrastructure. We have discussed some of the associated issues in previous chapters, such as the European Union Super Grid. The cost of transforming the energy infrastructure is a significant factor in determining the rate or transition to a new energy mix. An attempt to move too quickly away from fossil fuels can be disruptive to modern economies, while an attempt to move too slowly toward a sustainable energy mix can lead to geopolitical conflict over dwindling resources, and may lead to irreversible and undesirable changes to the climate. Government policy should seek to optimize the rate of transition.

We cannot say what the 21st century energy mix will be. We can say that the 21st century energy mix will depend on technological advances, including some advances that cannot be anticipated, and on choices made by society.

14.7 Activities

True-False

Specify if each of the following statements is True (T) or False (F).

1. Differences between forecasts and actual data can be used to test the validity of assumptions that are used to generate forecasts.
2. Forecasts based on the continuous extrapolation of historical trends can be invalidated by catastrophic events.
3. Forecasts of oil production may be based on the supply of oil or the demand for oil.
4. The United Nations Human Development Index is a measure of quality of life.
5. Sustainable development is a policy that encourages society to meet its current needs without regard for future needs.
6. Forecasts based on curve fitting work best when the inflection point (or peak) has been passed.
7. Most of the global energy infrastructure is designed to use renewable energy.
8. Analyses of historical data using a Gaussian curve typically predict that world oil production will peak after 2050.
9. Global future energy demand is expected to decline.
10. Forecasts of future energy production will change significantly when nuclear fusion becomes commercially viable.

Questions

1. What is energy conversion efficiency?
2. List three factors that are motivating a change in the energy mix.
3. What are the five objective criteria for evaluating an energy source used by Hodgson to develop a scenario in which the world would come to rely on nuclear fission energy?

4. What was an important environmental objective of Geller's [2003] renewable energy scenario?
5. List Schollnberger's [1999] three forecast scenarios.
6. Was Schollnberger's forecast of energy consumption based on supply or demand?
7. Was Hubbert's forecast of annual U.S. oil production based on supply or demand?
8. Is the cost of transforming the energy infrastructure a significant factor in determining the rate of transition from fossil fuels to a sustainable energy mix?
9. What factors should be considered in determining the rate of transition from fossil fuels to a sustainable energy mix?
10. Should government policy seek to optimize the rate of transition from fossil fuels to sustainable energy? Explain your answer.

APPENDIX A

UNITS AND SCIENTIFIC NOTATION

Typical units for expressing the amount of an energy source include barrels of oil, standard cubic feet of gas, tons of coal, and kilowatt-hours of electricity. Energy sources can be compared by converting to a common energy basis. Energy may be expressed in many different units, ranging from the British Thermal Unit (BTU) in the English system to the Joule (J) in SI (*System Internationale*) units. For example, the energy content of food is often expressed in Calories. One food Calorie is equivalent to 1000 calories or 4.184×10^3 J. Another important energy unit is the quad. A quad is a unit of energy that is often used in discussions of global energy because it is comparable in value to global energy values. One quad equals one quadrillion BTU (10^{15} BTU or approximately 10^{18} J).

SI Base Units

The SI base units and their symbols are presented in Table A-1. All other units can be derived from SI base units.

Table A-1
SI Base Units

Physical Quantity	Base Unit	Symbol
Length	meter	m
Mass	kilogram	kg
Time	second	s
Electric current	ampere	A
Temperature	kelvin	K
Amount of substance	mole	mol
Luminous intensity	candela	cd

An example of a derived unit is the Joule. The Joule is the SI unit of energy. It is a derived quantity with 1 J = 1 kg·m^2/s^2. The Watt is the SI unit of power. One Watt is equal to one Joule per second. One Watt/m^2 is equal to one Joule of energy passing through one square meter in one second. The unit Watt/m^2 can be used to compare different energy types. Table A-2 presents SI units for fundamental physical quantities.

Table A-2
SI Units for Fundamental Physical Quantities

Physical Quantity	Unit	Abbreviation	Comment
Length	meter	m	
Mass	kilogram	kg	
Time	second	s	
Force	Newton	N	1 N = 1 kg·m/s^2
Pressure	Pascal	Pa	1 Pa = 1 N/ m^2
Energy	Joule	J	1 J = 1 kg·m^2/s^2
Power	Watt	W	1 W = 1 J/s
Temperature	kelvin	K	
Amount of substance	mole	mol	
Frequency	Hertz	Hz	1 Hz = 1 cycle/s
Electric charge	Coulomb	C	
Electric current	Ampere	A	1 A = 1 C/s
Electric potential	Volt	V	1 V = 1J/C =1 W/A
Resistance	Ohm	Ω	1 Ω = 1 V/A
Capacitance	Farad	F	1 F = 1 C/V
Inductance	Henry	H	1 H = 1 V·s/A
Magnetic Induction $\vec{B}$	Tesla	T	1 T = 1 N/(A·m)
Magnetic flux	Weber	Wb	1 Wb = 1 T·m^2 = 1 V·s
Luminous intensity	candela	cd	

Scientific Notation

The physical quantities of interest here have values that range from very small to very large. Scientific notation is used to express the values of many physical quantities encountered in the study of energy. The most common prefixes for powers of 10 used in SI are presented in Table A-3.

Table A-3
Powers of 10

Prefix	Symbol	Value		Prefix	Symbol	Value
atto	A	10^{-18}		kilo	k	10^{3}
femto	F	10^{-15}		mega	M	10^{6}
pico	P	10^{-12}		giga	G	10^{9}
nano	N	10^{-9}		tera	T	10^{12}
micro	μ	10^{-6}		peta	P	10^{15}
milli	M	10^{-3}		exa	E	10^{18}

APPENDIX B

SUMMARY OF UNITED STATES HISTORICAL ENERGY PRODUCTION AND CONSUMPTION 1950-2008

Table B-1
U.S. Energy Consumption by Energy Source, 1950-2008
(Quadrillion BTU)

Energy Source	1950	1960	1970	1980	1990	2000	2008
Total	34.61	45.09	67.84	78.12	84.65	98.97	99.30
Fossil Fuels	31.63	42.14	63.52	69.83	72.33	84.73	83.44
Electricity Net Imports	0.01	0.02	0.01	0.07	0.01	0.12	0.11
Nuclear Electric Power	0.00	0.01	0.24	2.74	6.10	7.86	8.46
Renewable Energy	2.98	2.93	4.08	5.49	6.21	6.26	7.30
Fossil Fuel Source	**1950**	**1960**	**1970**	**1980**	**1990**	**2000**	**2008**
Total Fossil Fuels	31.63	42.14	63.52	69.83	72.33	84.73	83.44
Coal	12.35	9.84	12.26	15.42	19.17	22.58	22.42

Table B-1
U.S. Energy Consumption by Energy Source, 1950-2008
(Quadrillion BTU)

<table>
<tr><td>Coal Coke Net Imports</td><td>0.00</td><td>-0.01</td><td>-0.06</td><td>-0.04</td><td>0.01</td><td>0.07</td><td>0.04</td></tr>
<tr><td>Natural Gas[1]</td><td>5.97</td><td>12.39</td><td>21.80</td><td>20.24</td><td>19.60</td><td>23.82</td><td>23.84</td></tr>
<tr><td>Petroleum[2]</td><td>13.32</td><td>19.92</td><td>29.52</td><td>34.20</td><td>33.55</td><td>38.26</td><td>37.14</td></tr>
<tr><th>Renewable Source</th><th>1950</th><th>1960</th><th>1970</th><th>1980</th><th>1990</th><th>2000</th><th>2008</th></tr>
<tr><td>Total Renewable</td><td>2.98</td><td>2.93</td><td>4.08</td><td>5.49</td><td>6.21</td><td>6.26</td><td>7.30</td></tr>
<tr><td>Hydroelectric Conventional</td><td>1.42</td><td>1.61</td><td>2.63</td><td>2.90</td><td>3.05</td><td>2.81</td><td>2.45</td></tr>
<tr><td>Geothermal Energy</td><td>0.00</td><td>0.00</td><td>0.01</td><td>0.11</td><td>0.34</td><td>0.32</td><td>0.36</td></tr>
<tr><td>Solar/PV Energy</td><td>0.00</td><td>0.00</td><td>0.00</td><td>0.00</td><td>0.06</td><td>0.07</td><td>0.09</td></tr>
<tr><td>Wind Energy</td><td>0.00</td><td>0.00</td><td>0.00</td><td>0.00</td><td>0.03</td><td>0.06</td><td>0.51</td></tr>
<tr><td>Biomass</td><td>1.56</td><td>1.32</td><td>1.43</td><td>2.48</td><td>2.74</td><td>3.01</td><td>3.88</td></tr>
<tr><td colspan="8">1. Includes supplemental gaseous fuels.</td></tr>
<tr><td colspan="8">2. Petroleum products supplied, including natural gas plant liquids and crude oil burned as fuel.</td></tr>
<tr><td colspan="8">3. PV = photovoltaic</td></tr>
<tr><td colspan="8">4. Includes biofuels, waste (landfill gas, municipal solid waste (MSW) biofuels, etc.), wood and wood derived fuels.</td></tr>
</table>

Table B-2
U.S. Energy Production by Energy Source, 1950-2008
(Quadrillion BTU)

Energy Source	**1950**	**1960**	**1970**	**1980**	**1990**	**2000**	**2008**
Total	35.54	42.80	63.50	67.23	70.87	71.49	73.71
Fossil Fuels	32.56	39.87	59.19	59.01	58.56	57.37	57.94
Nuclear Electric Power	0.00	0.01	0.24	2.74	6.10	7.86	8.46
Renewable Energy	2.98	2.93	4.08	5.49	6.21	6.26	7.32
Fossil Fuel Source	**1950**	**1960**	**1970**	**1980**	**1990**	**2000**	**2008**
Total Fossil Fuels	32.56	39.87	59.19	59.01	58.56	57.37	57.94
Coal	14.06	10.82	14.61	18.60	22.49	22.74	23.86
Natural Gas (Dry)	6.23	12.66	21.67	19.91	18.33	19.66	21.15
Crude Oil	11.45	14.94	20.40	18.25	15.57	12.36	10.52
NGPL	0.82	1.46	2.51	2.25	2.18	2.61	2.42
Renewable Source	**1950**	**1960**	**1970**	**1980**	**1990**	**2000**	**2008**
Total Renewable	2.98	2.93	4.08	5.49	6.21	6.26	7.32
Hydroelectric Conventional	1.42	1.61	2.63	2.90	3.05	2.81	2.45
Geothermal Energy	0.00	0.00	0.01	0.11	0.34	0.32	0.36

Table B-2
U.S. Energy Production by Energy Source, 1950-2008
(Quadrillion BTU)

Solar/PV Energy	0.00	0.00	0.00	0.00	0.06	0.07	0.09
Wind Energy	0.00	0.00	0.00	0.00	0.03	0.06	0.51
Biomass	1.56	1.32	1.43	2.48	2.74	3.01	3.90
1. NGPL = natural gas plant liquids							

Table B-3
U.S. Energy Consumption by Energy Source, 1950-2008
(% of annual)

Energy Source	1950	1960	1970	1980	1990	2000	2008
Total	100.0	100.0	100.0	100.0	100.0	100.0	100.0
Fossil Fuels	91.4	93.5	93.6	89.4	85.4	85.6	84.0
Electricity Net Imports	0.0	0.0	0.0	0.1	0.0	0.1	0.1
Nuclear Electric Power	0.0	0.0	0.4	3.5	7.2	7.9	8.5
Renewable Energy	8.6	6.5	6.0	7.0	7.3	6.3	7.4
Fossil Fuel Source	**1950**	**1960**	**1970**	**1980**	**1990**	**2000**	**2008**
Total Fossil Fuels	100.0	100.0	100.0	100.0	100.0	100.0	100.0
Coal	39.0	23.3	19.3	22.1	26.5	26.6	26.9
Coal Coke Net Imports	0.0	0.0	-0.1	-0.1	0.0	0.1	0.0
Natural Gas[1]	18.9	29.4	34.3	29.0	27.1	28.1	28.6
Petroleum[2]	42.1	47.3	46.5	49.0	46.4	45.2	44.5
Renewable Source	**1950**	**1960**	**1970**	**1980**	**1990**	**2000**	**2008**
Total Renewable	100.0	100.0	100.0	100.0	100.0	100.0	100.0
Hydroelectric Conventional	47.5	54.9	64.6	52.9	49.1	44.9	33.6

Table B-3
U.S. Energy Consumption by Energy Source, 1950-2008
(% of annual)

Geothermal Energy	0.0	0.0	0.3	2.0	5.4	5.1	4.9
Solar/PV Energy	0.0	0.0	0.0	0.0	1.0	1.1	1.2
Wind Energy	0.0	0.0	0.0	0.0	0.5	0.9	7.0
Biomass	52.5	45.1	35.1	45.1	44.1	48.1	53.2
1. Includes supplemental gaseous fuels.							
2. Petroleum products supplied, including natural gas plant liquids and crude oil burned as fuel.							
3. PV = photovoltaic							
4. Includes biofuels, waste (landfill gas, municipal solid waste (MSW) biofuels, etc.), wood and wood derived fuels.							

Table B-4
U.S. Energy Production by Energy Source, 1950-2008
(% of annual)

Energy Source	1950	1960	1970	1980	1990	2000	2008
Total	100.0	100.0	100.0	100.0	100.0	100.0	100.0
Fossil Fuels	91.6	93.1	93.2	87.8	82.6	80.2	78.6
Nuclear Electric Power	0.0	0.0	0.4	4.1	8.6	11.0	11.5
Renewable Energy	8.4	6.8	6.4	8.2	8.8	8.8	9.9
Fossil Fuel Source	**1950**	**1960**	**1970**	**1980**	**1990**	**2000**	**2008**
Total Fossil Fuels	100.0	100.0	100.0	100.0	100.0	100.0	100.0
Coal	43.2	27.1	24.7	31.5	38.4	39.6	41.2
Natural Gas (Dry)	19.1	31.7	36.6	33.7	31.3	34.3	36.5
Crude Oil	35.2	37.5	34.5	30.9	26.6	21.5	18.2
NGPL	2.5	3.7	4.2	3.8	3.7	4.6	4.2
Renewable Source	**1950**	**1960**	**1970**	**1980**	**1990**	**2000**	**2008**
Total Renewable	100.0	100.0	100.0	100.0	100.0	100.0	100.0
Hydroelectric Conventional	47.5	54.9	64.6	52.9	49.1	44.9	33.5
Geothermal Energy	0.0	0.0	0.3	2.0	5.4	5.1	4.9

Table B-4
U.S. Energy Production by Energy Source, 1950-2008
(% of annual)

Solar/PV Energy	0.0	0.0	0.0	0.0	1.0	1.1	1.2
Wind Energy	0.0	0.0	0.0	0.0	0.5	0.9	7.0
Biomass	52.5	45.1	35.1	45.1	44.1	48.1	53.3
1. NGPL = natural gas plant liquids							

APPENDIX C

SUMMARY OF WORLD HISTORICAL ENERGY PRODUCTION AND CONSUMPTION 1970-2006

Table C-1
World Energy Consumption by Energy Source, 1980-2006
(Quadrillion BTU)

Energy Source	1970	1980	1990	2000	2006
Total	NA	283.3	347.7	397.9	472.3
Fossil Fuels		254.8	301.0	340.1	407.3
Nuclear Electric Power		7.6	20.4	25.7	27.8
Renewable Energy		20.8	26.3	32.2	37.3
Fossil Fuel Source	**1970**	**1980**	**1990**	**2000**	**2006**
Total Fossil Fuels	NA	254.8	301.0	340.1	407.3
Coal		70.0	89.2	93.6	127.6
Natural Gas		53.8	75.3	91.0	108.0
Petroleum Liquid		131.0	136.4	155.5	171.7
Renewable Source	**1970**	**1980**	**1990**	**2000**	**2006**
Total Renewable	NA	20.8	26.3	32.2	37.3
Hydroelectric Power		17.9	22.4	26.8	29.7
Geothermal, etc.		2.9	4.0	5.4	7.5

Table C-2
World Energy Production by Energy Source, 1970-2006
(Quadrillion BTU)

Energy Source	1970	1980	1990	2000	2006
Total	215.4	287.6	349.9	395.7	469.4
Fossil Fuels	200.8	259.1	303.3	337.9	404.5
Nuclear Electric Power	0.9	7.6	20.3	25.7	27.8
Renewable Energy	13.7	20.8	26.3	32.2	37.2
Fossil Fuel Source	**1970**	**1980**	**1990**	**2000**	**2006**
Total Fossil Fuels	200.8	259.1	303.3	337.9	404.5
Coal	63.0	71.3	91.0	90.4	128.5
Natural Gas	37.1	54.7	76.1	91.0	107.2
Crude Oil	97.1	128.0	129.4	146.8	157.1
NGPL[1]	3.6	5.1	6.9	9.6	11.7
Renewable Source	**1970**	**1980**	**1990**	**2000**	**2006**
Total Renewable	13.7	20.8	26.3	32.2	37.2
Hydroelectric Power	12.2	17.9	22.4	26.8	29.7
Geothermal, etc.	1.6	2.9	4.0	5.4	7.5

1. NGPL = natural gas plant liquids

Table C-3
World Energy Consumption by Energy Source, 1980-2006
(% of annual)

Energy Source	**1970**	**1980**	**1990**	**2000**	**2006**
Total	NA	100.0	100.0	100.0	100.0
Fossil Fuels		90.0	86.6	85.5	86.2
Nuclear Electric Power		2.7	5.9	6.4	5.9
Renewable Energy		7.4	7.6	8.1	7.9
Fossil Fuel Source	**1970**	**1980**	**1990**	**2000**	**2006**
Total Fossil Fuels	NA	100.0	100.0	100.0	100.0
Coal		27.5	29.6	27.5	31.3
Natural Gas		21.1	25.0	26.8	26.5
Petroleum Liquid		51.4	45.3	45.7	42.2
Renewable Source	**1970**	**1980**	**1990**	**2000**	**2006**
Total Renewable	NA	100.0	100.0	100.0	100.0
Hydroelectric Power		85.9	84.9	83.2	79.8
Geothermal, etc.		14.1	15.1	16.8	20.2

Table C-4
World Energy Production by Energy Source, 1970-2006
(% of annual)

Energy Source	1970	1980	1990	2000	2006
Total	100.0	100.0	100.0	100.0	100.0
Fossil Fuels	93.2	90.1	86.7	85.4	86.2
Nuclear Electric Power	0.4	2.6	5.8	6.5	5.9
Renewable Energy	6.4	7.2	7.5	8.1	7.9
Fossil Fuel Source	**1970**	**1980**	**1990**	**2000**	**2006**
Total Fossil Fuels	100.0	100.0	100.0	100.0	100.0
Coal	31.4	27.5	30.0	26.8	31.8
Natural Gas	18.5	21.1	25.1	26.9	26.5
Crude Oil	48.4	49.4	42.6	43.5	38.8
NGPL[1]	1.8	2.0	2.3	2.9	2.9
Renewable Source	**1970**	**1980**	**1990**	**2000**	**2006**
Total Renewable	100.0	100.0	100.0	100.0	100.0
Hydroelectric Power	88.4	85.9	84.9	83.2	79.9
Geothermal, etc.	11.6	14.1	15.1	16.8	20.1

1. NGPL = natural gas plant liquids

REFERENCES

Abengoa Solar, 2009, http://www.abengoasolar.com/, accessed December 30, 2009.

Aubrecht, G.J., 1995, **Energy**, 2nd Edition, Prentice-Hall, Inc., Upper Saddle River, New Jersey.

Ausubel, J.H., 2000, "Where is Energy Going?" *The Industrial Physicist* (February), pages 16-19.

AWEA, 2009, American Wind Energy Association, http://www.awea.org/, accessed December 2009.

Bain, A., and W.D. Van Vorst, 1999, "The Hindenburg tragedy revisited: the fatal flaw found," *International Journal of Hydrogen Energy*, Volume 24, pages 399-403.

Barker, S., 2009a, "More Suits Against TVA," knoxnews.com, posted November 27, 2009, accessed December 5, 2009.

Barker, S., 2009b, "Panel Calls for Tougher Regulation of Ash Sites," knoxnews.com, posted December 2, 2009, accessed December 5, 2009.

Bartlett, A.A., 2004, "Thoughts on Long-Term Energy Supplies: Scientists and the Silent Lie," *Physics Today* (July), pages 53-55.

Borbely, A. and J.F. Kreider, 2001, **Distributed Generation: The Power Paradigm for the New Millenium**, CRC Press, New York, New York.

Boswell, R., 2009, "Is Gas Hydrate Energy Within Reach?" *Science* (21 August), pages 957-958.

Brennan, T.J., K.L. Palmer, R.J. Kopp, A.J. Krupnick, V. Stagliano, and D. Burtraw, 1996, **A Shock to the System: Restructuring America's Electricity Industry**, Resources for the Future, Washington, D.C.

Campbell, C.J., 2005, "Just how much oil does the Middle East really have, and does it matter?" *Oil & Gas Journal* (4 April), pages 24-26.

Carbon, M.W., 1997, **Nuclear Power: Villain or Victim?**, Pebble Beach Publishers, Madison, Wisconsin.

Cassedy, E.S. and P.Z. Grossman, 1998, **Introduction to Energy**, 2nd Edition, Cambridge U.P., Cambridge, U.K..

Club of Rome, 2010, http://www.clubofrome.org/, accessed January 1, 2010.

Cook, E., 1971, "The Flow of Energy in an Industrial Society," *Scientific American* (September), pages 135-144.

Crotogino, F, K-U. Mohmeyer, and R. Scharf, 2001, Huntorf CAES: More than 20 Years of Successful Operation, http://www.unisaarland.de/fak7/fze/AKE_Archiv/AKE2003H/AKE2003H_Votraege/AKE2003H03c_Crotogino_ea_HuntorfCAES_CompressedAirEnergyStorage.pdf, accessed December 30, 2009.

Crabtree, G.W. and N.S. Lewis, 2007, "Solar Energy Conversion," *Physics Today* (March), pages 37-42.

Crowe, B.J., 1973, *Fuel Cells – A Survey*, NASA, U.S. Government Printing Office, Washington, D.C.

CRU website, 2010, Climatic Research Unit, University of East Anglia, United Kingdom, http://www.cru.uea.ac.uk/cru/info/warming/, accessed January 6, 2010, original article http://www.agu.org/pubs/crossref/2006/2005JD006548.shtml.

CTGP, 2009, China Three Gorges Project, http://www.ctgpc.com/, accessed December 27, 2009.

Dahl, C., 2004, **International Energy Markets**, PennWell Books, Tulsa, Oklahoma.

Dawson, J., 2002, "Fusion Energy Panel Urges US to Rejoin ITER," *Physics Today* (November), pages 28-29.

Deffeyes, K.S., 2001, **Hubbert's Peak – The Impending World Oil Shortage**, Princeton U.P., Princeton, New Jersey.

Desertec Red Paper, 2010, *An Overview of the Desertec Project, 2nd Edition*, The Desertec Foundation, an initiative of the Club of Rome, http://www.desertec.org/fileadmin/downloads/, accessed January 1, 2010.

Desertec White Paper, 2010, *Clean Power from Deserts*, The Desertec Foundation, an initiative of the Club of Rome, http://www.desertec.org/fileadmin/downloads/, accessed January 1, 2010.

DoE Geothermal, 2002, "Geothermal Energy Basics," United States Department of Energy, http://www.eren.doe.gov/geothermal/geobasics.html, accessed October 23, 2002.

DoE Hydropower, 2002, "Hydropower," United States Department of Energy, http://www.eren.doe.gov/RE/hydropower, accessed October 24, 2002.

DoE Hydropower History, 2009, "Hydropower," United States Department of Energy, http://www.eren.doe.gov/RE/hydropower, accessed December 27, 2009.

DoE Ocean, 2002, "Ocean," United States Department of Energy, http://www.eren.doe.gov/RE/Ocean.html, accessed October 24, 2002.

DoE Oklo, 2009, "Oklo: Natural Nuclear Reactors," United States Department of Energy Office of Civilian Radioactive Waste Management, http://www.ocrwm.doe.gov/factsheets/doeymp0010.shtml, accessed December 29, 2009.

DoE Power Grid, 2010, Distributed Energy Program, United States Department of Energy, http://www.eere.energy.gov/de/us_power_grids.html, accessed January 1, 2010.

ENTSO-E website, 2010, European Network of Transmission System Operators for Electricity, http://www.entsoe.eu/, accessed January 1, 2010.

EWEA, 2009, *Pure Power 2009*, European Wind Energy Association Report, 2009 Update, http://www.ewea.org/index.php?id=178, accessed December 27, 2009.

Fanchi, J.R., 2004, **Energy: Technology and Directions for the Future**, Elsevier-Academic Press, Boston.

Fanchi, J.R., 2010, **Integrated Reservoir Asset Management**, Elsevier-Academic Press, Boston.

Feder, T., 2009, "JET gets new wall to prep for ITER," *Physics Today*, December 2009, pages 24-25.

Geller, H., 2003, **Energy Revolution**, Island Press, Washington.

Gold, T., 1999, **The Deep Hot Biosphere**, Springer-Verlag New York, Inc., New York, New York.

Goldemberg, J., 2008, "The Brazilian biofuels industry," *Biotechnology for Biofuels*, published 1 May 2008, doi:10.1186/1754-6834-1-6, http://www.biotechnologyforbiofuels.com/content/1/1/6, accessed January 2010.

Goswami, D.Y., F. Kreith, and J.F. Kreider, 2000, **Principles of Solar Engineering**, George H. Buchanan Co., Philadelphia, Pennsylvania.

Graw, K-U., 2008, Wiley-VCH, "Energy Reserves from the Oceans," **Renewable Energy**, Edited by G. Wengenmayr and T. Buhrke, editors, pages 76-82.

Hayden, H.C., 2001, **The Solar Fraud: Why Solar Energy Won't Run the World**, Vales Lake Publishing, LLC, Pueblo West, Colorado.

Hodgson, P.E., 1999, **Nuclear Power, Energy and the Environment**, Imperial College Press, London, United Kingdom.

Hubbert, M.K., 1956, "Nuclear Energy and the Fossil Fuels," American Petroleum Institute Drilling and Production Practice, Proceedings of the Spring Meeting, San Antonio, pages 7-25.

Huntington, S.P., 1996, **The Clash of Civilizations**, Simon and Schuster, London.

IER website, 2010, "Levelized Cost of New Electricity Generating Technologies," study posted 12 May 2009, by the Institute for Energy Research, http://www.instituteforenergyresearch.org/, accessed January 4, 2010.

International Rivers, 2009, "China's Three Gorges Dam," http://www.internationalrivers.org/en/china/three-gorges-dam, accessed December 27, 2009.

IPCC, 2007, IPCC Climate Change 2007 report, Intergovernmental Panel on Climate Change, http://www.ipcc.ch/publications_and_data/ar4/wg1/en/faq-2-1-figure-1.html, accessed January 8, 2010.

ITER, 2009, International Thermonuclear Experimental Reactor, http://www.iter.org/default.aspx, accessed December 22, 2009.

Jenkins, C.D., Jr., and C.M. Boyer II, 2008, "Coalbed- and Shale-Gas Reservoirs," *Journal of Petroleum Technology*, pages 92-99 (February).

Klare, M.T., 2001, **Resource Wars: The New Landscape of Global Conflict**, Henry Holt and Company, New York.

Klare, M.T., 2004, **Blood and Oil**, Henry Holt and Company, New York.

Kühn, M., 2008a, Wiley-VCH, "A Tailwind for Sustainable Technology," *Renewable Energy*, Edited by G. Wengenmayr and T. Buhrke, editors, pages 14-20.

Kraushaar, J.J. and R.A. Ristinen, 1993, **Energy and Problems of a Technical Society**, 2nd Edition, Wiley, New York.

Laherrère, J.H., 2000, "Learn strengths, weaknesses to understand Hubbert curves," *Oil and Gas Journal*, pages 63-76 (17 April); see also Laherrère's earlier article "World oil supply – what goes up must come down, but when will it peak?" *Oil and Gas Journal*, pages 57-64 (1 February 1999) and letters in *Oil and Gas Journal*, (1 March 1999).

Lide, D.R., 2002, **CRC Handbook of Chemistry and Physics**, 83rd Edition, CRC Press, Boca Raton, Florida.

Lilley, J., 2001, **Nuclear Physics**, Wiley, New York.

Lovelock, J., 2009, **The Vanishing Face of GAIA**, Basic Books, New York.

Manwell, J.F., J.G. McGowan, and A.L. Rogers, 2002, **Wind Energy Explained**, Wiley, New York.

Marcum, E., 2009, "TVA Realigns Top Management," knoxnews.com, posted December 5, 2009, accessed December 5, 2009.

Mathews, J.N.A., 2009, "Superconductors to boost wind power," *Physics Today* (April), pages 25-26.

McKay, D.R. (correspondent), 2002, "Global Scenarios 1998-2020," Summary Brochure, Shell International, London.

McVeigh, J.C., 1984, **Energy Around the World**, Pergamon Press, Oxford, United Kingdom.

Mills, R., 2009, "The Myth of the Oil Crisis," *Journal of Petroleum Technology*, pages 16-17 (October).

Morrison, P. and K. Tsipis, 1998, **Reason Enough to Hope**, The MIT Press, Cambridge, Massachusetts, especially Chapter 9.

Murray, R.L., 2001, **Nuclear Energy: An Introduction to the Concepts, Systems, and Applications of Nuclear Processes**, 5th Edition, Butterworth-Heinemann, Boston, Massachusetts.

Nef, J.U., 1977, "An Early Energy Crisis and its Consequences," *Scientific American* (November), pages 140-151.

NASA Earth Observatory, 2009, http://earthobservatory.nasa.gov/, accessed December 30, 2009.

NGSA Unconventional Gas, National Gas Supply Association, http://www.naturalgas.org/overview/, accessed December 20, 2009.

NWA, 2001, "Radioactive Wastes," Nuclear Waste Association, http://www.world-nuclear.org/info/inf60.html, accessed December 29, 2009.

Ogden, J.M., 2002, "Hydrogen: The Fuel of the Future?" *Physics Today* (April), pages 69-75.

Pitz-Paal, R., 2008, Wiley-VCH, "How the Sun gets into the Power Plant," **Renewable Energy**, Edited by G. Wengenmayr and T. Buhrke, editors, pages 26-33.

Plocek, T.J., M. Laboy, and J.A. Marti, 2009, "Ocean Thermal Energy Conversion (OTEC): Technical Viability, Cost Projections and Development Strategies," Paper OTC 19979, proceedings of the 2009 Offshore Technology Conference, Houston, May 4-7, 2009.

Poupee, K., 2009, "Japan eyes solar power station in space," Discovery News, http://news.discovery.com/space/japan-solar-space-station.html, posted November 9, 2009; accessed December 23, 2009.

Ramage, J. and J. Scurlock, 1996, "Biomass," **Renewable Energy: Power for a Sustainable Future**, edited by G. Boyle, Oxford University Press, Oxford, United Kingdom.

Ristinen, R.A. and J.J. Kraushaar, 1999, **Energy and the Environment**, Wiley, New York.

Schollnberger, W.E., 1999, "Projection of the World's Hydrocarbon Resources and Reserve Depletion in the 21st Century," *The Leading Edge* (May), pages 622-625.

Schollnberger, W.E., 2006, "Who Shapes the Future Mix of Primary Energy? What Might it Be?" *OIL GAS European Magazine*, Volume 32, Number 1, pages 8-20.

Serway, R.A. and J.S. Faughn, 1985, **College Physics**, Saunders, Philadelphia.

Shepherd, W. and D.W. Shepherd, 1998, **Energy Studies**, Imperial College Press, London, U.K.

Silberberg, M., 1996, **Chemistry**, Mosby, St. Louis. Solar Spaces, 2009, http://www.solarpaces.org/Tasks/Task1/solar_tres.htm, accessed December 30, 2009.

Sørensen, B., 2000, **Renewable Energy: Its physics, engineering, environmental impacts, economics & planning**, 2nd Edition, Academic Press, London, U.K.

SPE Definitions, 2009, Development_of_Definitions.pdf, Society of Petroleum Engineers, http://www.spe.org/, accessed December 6, 2009.

SPE-PRMS, 2007, Society of Petroleum Engineers, http://www.spe.org/industry/reserves/docs/Petroleum_Resources_Management_System_2007.pdf, accessed December 6, 2009.

TVA, 2009, Tennessee Valley Authority, http://www.tva.gov/sites/kingston.htm, accessed December 6, 2009.

UN HDI, 2009, United Nations Human Development Index, http://hdr.undp.org/en/statistics/, accessed November 25, 2009.

US-Canada Blackout Report, 2004, "Final Report on the August 14, 2003 Blackout in the United States and Canada: Causes and Recommendations," U.S.-Canada Power System Outage Task Force, https://reports.energy.gov/BlackoutFinal-Web.pdf, accessed January 1, 2010.

US Census website, 2010, United States Census Bureau, http://www.census.gov/ipc/www/worldhis.html, accessed January 8, 2010.

US EIA website, 2001, Annual Energy Review 2001, Appendix F, United States Energy Information Agency, http://www.eia.doe.gov/.

US EIA website, 2002, Table 11.1, United States Energy Information Agency, http://www.eia.doe.gov/.

US EIA website, 2008, Annual Energy Review 2008, United States Energy Information Agency, http://www.eia.doe.gov/.

US EIA website, 2009, United States Energy Information Agency, http://www.eia.doe.gov/.

US EIA website, 2010, International Energy Statistics, United States Energy Information Agency, http://www.eia.doe.gov/.

US FERC website, 2010, United States Federal Energy Regulatory Commission, http://www.ferc.gov/industries/electric/indus-act/smart-grid.asp, accessed January 1, 2010.

US NOAA website, 2010, United States National Oceanic and Atmospheric Administration, http://www.noaanews.noaa.gov/stories/s2015.htm, accessed January 1, 2010.

USGS Hydrate, 2009, United States Geological Survey, http://geology.usgs.gov/connections/mms/joint_projects/methane.htm, accessed December 2009.

USGS website, 2009, United States Geological Survey, http://pubs.usgs.gov/gip/dynamic/historical.html, accessed December 2009.

van Dyke, K., 1997, *Fundamentals of Petroleum*, 4th Edition, Petroleum Extension Service, University of Texas, Austin.

Varian, H.R., 2009, **Intermediate Microeconomics**, 8th Edition, W.W. Norton and Company, New York.

Verleger, P.K. Jr., 2000, "Third Oil Shock: Real or Imaginary?" *Oil & Gas Journal* (12 June), pages 76-88.

WCED (World Commission on Environment and Development), Brundtland, G., Chairwoman, 1987, **Our Common Future**, Oxford University Press.

WCI, 2009, World Coal Institute, accessed December 15, 2009, http://www.worldcoal.org/coal/coal-seam-methane/coal-bed-methane/.

WEC, 2007, *2007 Survey of Energy Resources*, World Energy Council, data from mid-2005, http://www.worldenergy.org/, accessed December 20, 2009.

Weisz, P.B., 2004, "Basic Choices and Constraints on Long-Term Energy Supplies," *Physics Today* (July), pages 47-52.

Wengenmayr, G., 2008a, Wiley-VCH, "Flowing Energy," **Renewable Energy**, Edited by G. Wengenmayr and T. Buhrke, editors, pages 22-25.

Whittaker, M., 1999, "Emerging 'triple bottom line' model for industry weighs environmental, economic, and social considerations," *Oil and Gas Journal*, pages 23-28 (20 December).

Wigley, T.M.L., R. Richels, and J.A. Edmonds, 1996, "Economic and environmental choices in the stabilization of atmospheric CO2 concentrations," *Nature* (18 January), pages 240-243.

Wiser, W.H., 2000, **Energy Resources: Occurrence, Production, Conversion, Use**, Springer-Verlag New York, Inc., New York, New York.

WNA, March 2001, "Radioactive Wastes," World Nuclear Association, http://www.world-nuclear.org/, accessed December 30, 2009.

WNA, September 2009, "Supply of Uranium," World Nuclear Association, http://www.world-nuclear.org/, accessed December 30, 2009.

Yergin, D., 1992, **The Prize**, Simon and Schuster, New York.

INDEX

Q

R

S

T

U

V

W

Y